PHILIP'S

STARGAZING 2014

MONTH-BY-MONTH GUIDE TO THE NORTHERN NIGHT SKY

HEATHER COUPER & NIGEL HENBEST

www.philips-maps.co.uk

HEATHER COUPER and NIGEL HENBEST are internationally recognized writers and broadcasters on astronomy, space and science. They have written more than 40 books and over 1000 articles, and are the founders of an independent TV production company specializing in factual and scientific programming.

Heather is a past President of both the British Astronomical Association and the Society for Popular Astronomy. She is a Fellow of the Royal Astronomical Society, a Fellow of the Institute of Physics and a former Millennium Commissioner, for which she was awarded the CBE in 2007. Nigel has been Astronomy Consultant to *New Scientist* magazine, Editor of the *Journal of the British Astronomical Association* and Media Consultant to the Royal Greenwich Observatory.

Heather Couper

Nigel Henbest

Published in Great Britain in 2013
by Philip's,
a division of Octopus Publishing Group Limited
(www.octopusbooks.co.uk)
Endeavour House, 189 Shaftesbury Avenue,
London WC2H 8JY
An Hachette UK Company (www.hachette.co.uk)

Reprinted 2013

TEXT
Heather Couper and Nigel Henbest (pages 6–53)
Robin Scagell (pages 61–64)
Philip's (pages 1–5, 54–60)

ISBN 978–1–84907–290–8

Printed in Italy

Title page: *The Heart Nebula (James McConnachie/Galaxy)*

ACKNOWLEDGEMENTS

All star maps by Wil Tirion/Philip's, with extra annotation by Philip's. Artworks © Philip's.

All photographs courtesy of Galaxy Picture Library:
Alan Clitherow *52;*
Jim Coughlan *32;*
Nick Hart *24;*
Rob Hodgkinson *41;*
Tom Howard *36;*
James McConnachie *44;*
Damian Peach *8;*
Robin Scagell *28, 48, 61–64;*
Peter Shah *12;*
Dave Tyler *16, 20.*

CONTENTS

The sight of diamond-bright stars sparkling against a sky of black velvet is one of life's most glorious experiences. No wonder stargazing is so popular. Learning your way around the night sky requires nothing more than patience, a reasonably clear sky and the 12 star charts included in this book.

Stargazing 2014 is a guide to the sky for every month of the year. Complete beginners will use it as an essential night-time companion, while seasoned amateur astronomers will find the updates invaluable.

THE MONTHLY CHARTS

Each pair of monthly charts shows the views of the heavens looking north and south. They are usable throughout most of Europe – between 40 and 60 degrees north. Only the brightest stars are shown (otherwise we would have had to put 3000 stars on each chart, instead of about 200). This means that we plot stars down to third magnitude, with a few fourth-magnitude stars to complete distinctive patterns. We also show the ecliptic, which is the apparent path of the Sun in the sky.

USING THE STAR CHARTS

To use the charts, begin by locating the north Pole Star – Polaris – by using the stars of the Plough (see July). When you are looking at Polaris you are facing north, with west on your left and east on your right. (West and east are reversed on star charts because they show the view looking up into the sky instead of down towards the ground.) The left-hand chart then shows the view you have to the north. Most of the stars you see will be circumpolar, which means that they are visible all year. The other stars rise in the east and set in the west.

Now turn and face the opposite direction, south. This is the view that changes most during the course of the year. Leo, with its prominent 'sickle' formation, is high in the spring skies. Summer is dominated by the bright trio of Vega, Deneb and Altair. Autumn's familiar marker is the Square of Pegasus, while the winter sky is ruled over by the stars of Orion.

The charts show the sky as it appears in the late evening for each month: the exact times are noted in the caption with the chart. If you are observing in the early morning, you will find that the view is different. As a rule of thumb, if you are observing two hours later than the time suggested in the caption, then the following month's map will more accurately represent the stars on view. So, if you wish to observe at midnight in the middle of February, two hours later than the time suggested in the caption, then the stars will appear as they are on March's chart. When using a chart for the 'wrong' month, however, bear in mind that the planets and Moon will not be shown in their correct positions.

THE MOON, PLANETS AND SPECIAL EVENTS

In addition to the stars visible each month, the charts show the positions of any planets on view in the late evening. Other planets may also be visible that month, but they will not be on the chart if they have already set, or if they do not rise until early morning. Their positions are described in the text, so that you can find them if you are observing at other times.

We have also plotted the path of the Moon. Its position is marked at three-day intervals. The dates when it reaches First Quarter, Full Moon, Last Quarter and New Moon are given in the text. If there is a meteor shower in the month, we mark the position from which the meteors appear to emanate – the *radiant*. More information on observing the planets and other Solar System objects is given on pages 54–57.

Once you have identified the constellations and found the planets, you will want to know more about what's on view. Each month, we explain one object, such as a particularly interesting star or galaxy, in detail. We have also chosen a spectacular image for each month and described how it was captured. All of these pictures were taken by amateurs. We list details and dates of special events, such as meteor showers or eclipses, and give observing tips. Finally, each month we pick a topic related to what's on view, ranging from aurorae to star colours and extrasolar planets, and discuss it in more detail. Where possible, all relevant objects are highlighted on the maps.

FURTHER INFORMATION

The year's star charts form the heart of the book, providing material for many enjoyable observing sessions. For background information turn to pages 54–57, where diagrams help to explain, among other things, the movement of the planets and why we see eclipses.

Although there is plenty to see with the naked eye, many observers use binoculars or telescopes, and some choose to record their observations using cameras, CCDs or webcams. For a round-up of what's new in observing technology, go to pages 61–64, where equipment expert Robin Scagell shares his knowledge of remote observing.

If you have already invested in binoculars or a telescope, then you can explore the deep sky – nebulae (starbirth sites), star clusters and galaxies. On pages 58–60 we list recommended deep-sky objects, constellation by constellation. Use the appropriate month's maps to see which constellations are on view, and then choose your targets. The table of 'limiting magnitude' (page 58) will help you to decide if a particular object is visible with your equipment.

Happy stargazing!

This is a great month to get to know the planets. **Jupiter**, in particular, is resplendent. Then Venus, Mars, Saturn, **Uranus**, Neptune – and even little Mercury – grace our skies. And they're matched by the magnificent constellations of winter. These are so striking and recognizable that there's no better time to start finding your way around the heavens. The dazzling denizens of **Orion**, **Taurus**, **Gemini** and **Canis Major** make up a scintillating celestial tableau.

▼ The sky at 10 pm in mid-January, with Moon positions at three-day intervals either side of Full Moon. The star positions are also correct for 11 pm at

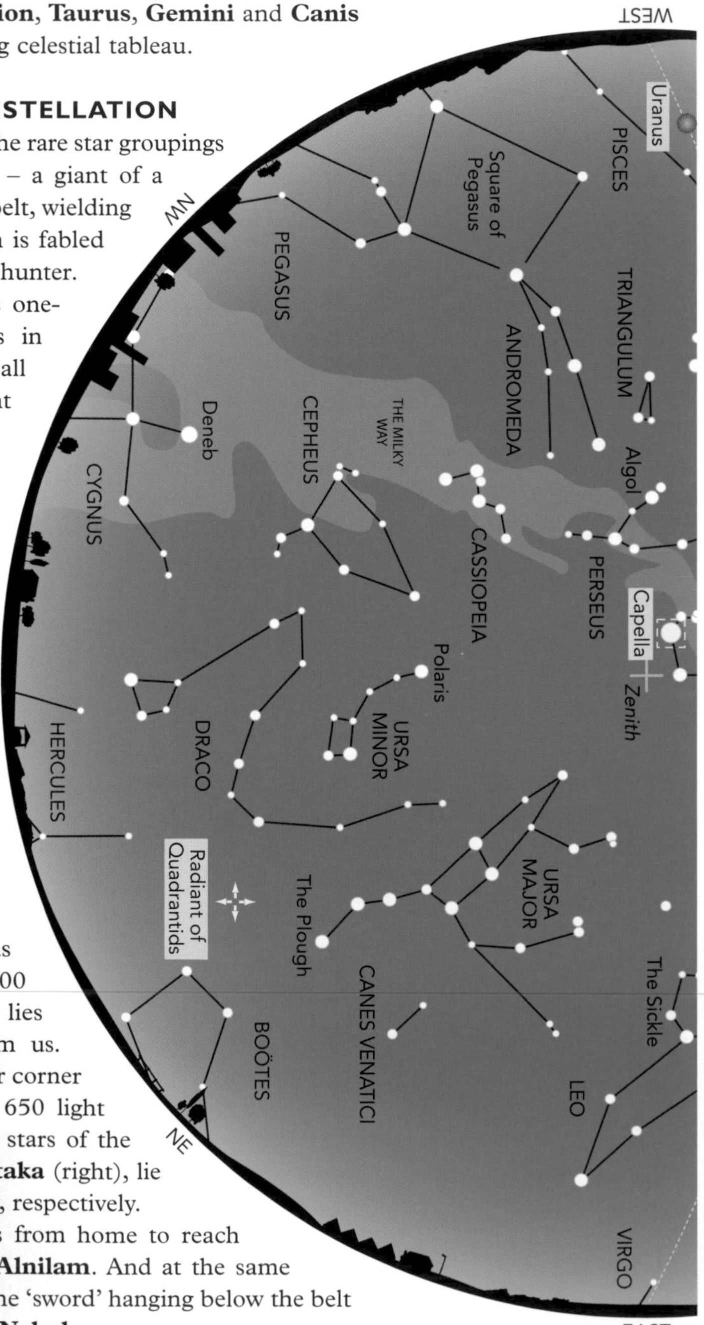

JANUARY'S CONSTELLATION

Spectacular **Orion** is one of the rare star groupings that looks like its namesake – a giant of a man with a sword below his belt, wielding a club above his head. Orion is fabled in mythology as the ultimate hunter.

The constellation contains one-tenth of the brightest stars in the sky: its seven main stars all lie in the 'top 70' of brilliant stars. Despite its distinctive shape, most of these stars are not closely associated with each other – they simply line up, one behind the other.

Closest is the star that forms the hunter's right shoulder, **Bellatrix,** at 250 light years. Next is blood-red **Betelgeuse** at the top left of Orion, 640 light years away.

The brightest star in the constellation, blue-white **Rigel**, is a vigorous young star more than twice as hot as our Sun, and nearly 120,000 times more luminous. Rigel lies around 860 light years from us. **Saiph,** which marks the other corner of Orion's tunic, is around 650 light years distant. The two outer stars of the belt, **Alnitak** (left) and **Mintaka** (right), lie 740 and 900 light years away, respectively.

We travel 1300 light years from home to reach the middle star of the belt, **Alnilam**. And at the same distance, we see the stars of the 'sword' hanging below the belt – the lair of the great **Orion Nebula**.

WEST
Uranus
PISCES
Square of Pegasus
PEGASUS
TRIANGULUM
ANDROMEDA
Deneb
CEPHEUS
THE MILKY WAY
Algol
CYGNUS
CASSIOPEIA
PERSEUS
Capella
Polaris
Zenith
NORTH
NW
HERCULES
DRACO
URSA MINOR
Radiant of Quadrantids
The Plough
URSA MAJOR
CANES VENATICI
BOÖTES
The Sickle
LEO
NE
VIRGO
EAST

the beginning of January, and 9 pm at the end of the month. The planets move slightly relative to the stars during the month.

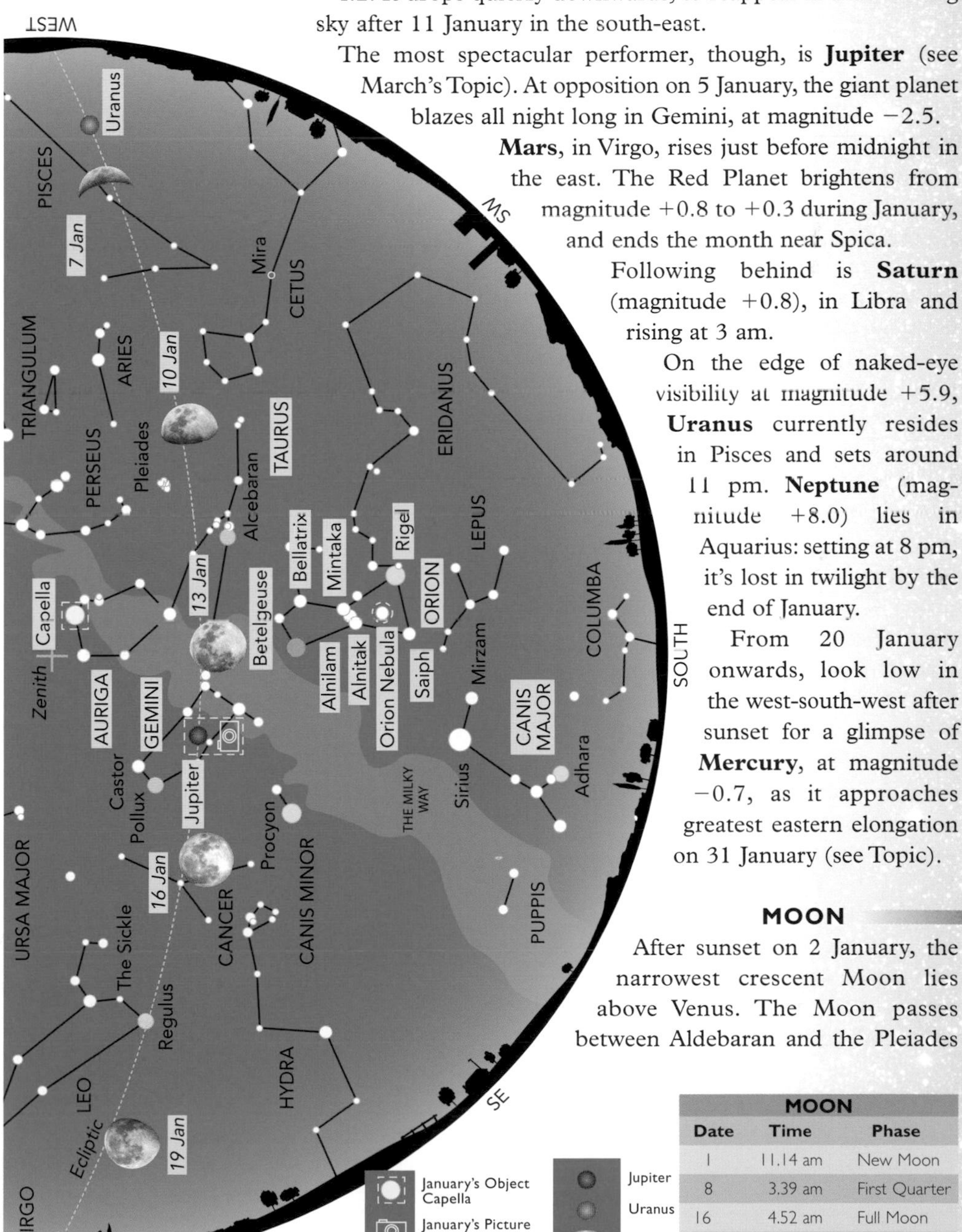

PLANETS ON VIEW

Venus puts on an entertaining act this month, appearing first as Evening Star and then as Morning Star – and, in between, as both at once (see Special Events)! At the start of the year, you'll spot Venus low in the south-west after sunset, at magnitude −4.2. It drops quickly downwards, to reappear in the morning sky after 11 January in the south-east.

The most spectacular performer, though, is **Jupiter** (see March's Topic). At opposition on 5 January, the giant planet blazes all night long in Gemini, at magnitude −2.5.

Mars, in Virgo, rises just before midnight in the east. The Red Planet brightens from magnitude +0.8 to +0.3 during January, and ends the month near Spica.

Following behind is **Saturn** (magnitude +0.8), in Libra and rising at 3 am.

On the edge of naked-eye visibility at magnitude +5.9, **Uranus** currently resides in Pisces and sets around 11 pm. **Neptune** (magnitude +8.0) lies in Aquarius: setting at 8 pm, it's lost in twilight by the end of January.

From 20 January onwards, look low in the west-south-west after sunset for a glimpse of **Mercury**, at magnitude −0.7, as it approaches greatest eastern elongation on 31 January (see Topic).

MOON

After sunset on 2 January, the narrowest crescent Moon lies above Venus. The Moon passes between Aldebaran and the Pleiades

MOON		
Date	**Time**	**Phase**
1	11.14 am	New Moon
8	3.39 am	First Quarter
16	4.52 am	Full Moon
24	5.19 am	Last Quarter
30	9.38 pm	New Moon

on 11 January. The brilliant planet Jupiter is near the Moon on 14 and 15 January. The Moon moves below Regulus on 18 January. On the night of 22/23 January, you'll find the Moon to the right of Mars and Spica. The morning of 25 January sees the crescent Moon next to Saturn; it's near Venus just before dawn on 28 and 29 January.

SPECIAL EVENTS

The night of **3/4 January** sees the maximum of the **Quadrantid** meteor shower. These shooting stars are tiny particles of dust shed by the old comet 2003 EH_1 that burn up – often in a blue or yellow streak – as they enter the Earth's atmosphere. This is an excellent year for viewing the Quadrantids, as the Moon sets early.

On **4 January**, at 11.59 am, the Earth is at perihelion, its closest point to the Sun – a 'mere' 147 million kilometres away.

On **11 January**, Venus is at inferior conjunction, appearing to fly 5 degrees over the Sun as it passes between the Earth and the Sun. It's a rare chance to see Venus as both an Evening Star and a Morning Star on the same night!

JANUARY'S OBJECT

Capella, 'the little she goat', sails right overhead in our winter skies. This yellow giant star – the brightest in the constellation of **Auriga** (the Charioteer) – is anything but diminutive. It's the sixth-brightest star in the sky, outshining our Sun almost 80 times. And there's more to Capella than meets the eye: it has a large yellow companion star in a close orbit about it (not visible through a small telescope). Further out, the system includes two faint red dwarfs. Capella and its major companion were probably once hot, white stars. But after gobbling up their reserves of nuclear fuel, these are stars on their way out. They will eventually become red giants – like Betelgeuse.

◀ Damian Peach used a Flea3 webcam-type camera with separate red, green and blue filters on his 355 mm Celestron Schmidt-Cassegrain telescope to capture this image of Jupiter. This is the inverted view through an astronomical telescope, with north at the bottom.

JANUARY'S PICTURE

This stunning portrait of the largest world in our Solar System – **Jupiter** – was taken by a British astronomer observing from Barbados, where sky conditions are excellent. The huge gas giant spins so rapidly that its clouds are drawn out into streaks by the speed of the planet's rotation. Jupiter is a delight to observe: its atmosphere is ever-changing. One constant is the Great Red Spot (top left) – a storm-system three times bigger than Earth. The dots either side of the planet are two of its brightest moons: Io (lower left) and Ganymede (top right).

Viewing tip

Don't think that you need a telescope to bring the heavens closer. Binoculars are excellent – and you can fling them into the back of the car at the last minute. But when you buy binoculars, make sure that you get those with the biggest lenses, coupled with a modest magnification. Binoculars are described, for instance, as being '7×50' – meaning that the magnification is seven times, and that the diameter of the lenses is 50 mm. These are ideal for astronomy – they have good light grasp, and the low magnification means that they don't exaggerate the wobbles of your arms too much. It's always best to rest your binoculars on a wall or a fence to steady the image. Some amateurs are the lucky owners of huge binoculars – say, 20×70 – with which you can see the rings of Saturn (being so large, these binoculars need a special mounting). But above all, *never* buy binoculars with small lenses that promise huge magnifications – they're a total waste of money.

JANUARY'S TOPIC
Messenger to the gods

Look out for hard-to-spot Mercury this month. Rumour has it that the architect of our Solar System – Nicolaus Copernicus – never observed the tiny world because of mists rising from the nearby River Vistula in Poland. Being the closest planet to the Sun, it seldom strays far from the glare of our local star. But at the end of January, Mercury puts on its best evening appearance of the year, setting two hours after the Sun.

The pioneering space probe Mariner 10 sent back only brief images as it swung past the diminutive, cratered planet in 1974. But everything changed in 2011, when NASA's Messenger probe went into orbit around Mercury.

Messenger (the convoluted acronym stands for ME-rcury S-urface S-pace EN-vironment GE-ochemistry and R-anging mission) celebrates the belief that, in mythology, fleet-footed Mercury was messenger to the gods. Messenger's instruments are scanning the planet's surface and scrutinizing its composition – offering clues to the origins of this mysterious body.

The spacecraft is also exploring Mercury's internal workings – and it's found this small world has a core made of molten iron. Plus: Messenger has discovered water in the planet's thin outer atmosphere, as well as evidence for past volcanic activity. Most exciting, peering inside permanently-shaded craters at the planet's north pole, Messenger has found signs of organic compounds and ice sheets up to 20 metres thick.

The first signs of spring are now on the way, as the winter star-patterns start to drift towards the west, setting earlier. The constantly-changing pageant of constellations in the sky is proof that we live on a cosmic carousel, orbiting the Sun. Imagine: you're in the fairground, circling the merry-go-round on your horse, and looking out around you. At times you spot the ghost train; sometimes you see the roller-coaster; and then you swing past the candy-floss stall. So it is with the sky – and the constellations – as we circle our local star. That's why we get to see different stars in different seasons.

▼ *The sky at 10 pm in mid-February, with Moon positions at three-day intervals either side of Full Moon. The star positions are also correct for*

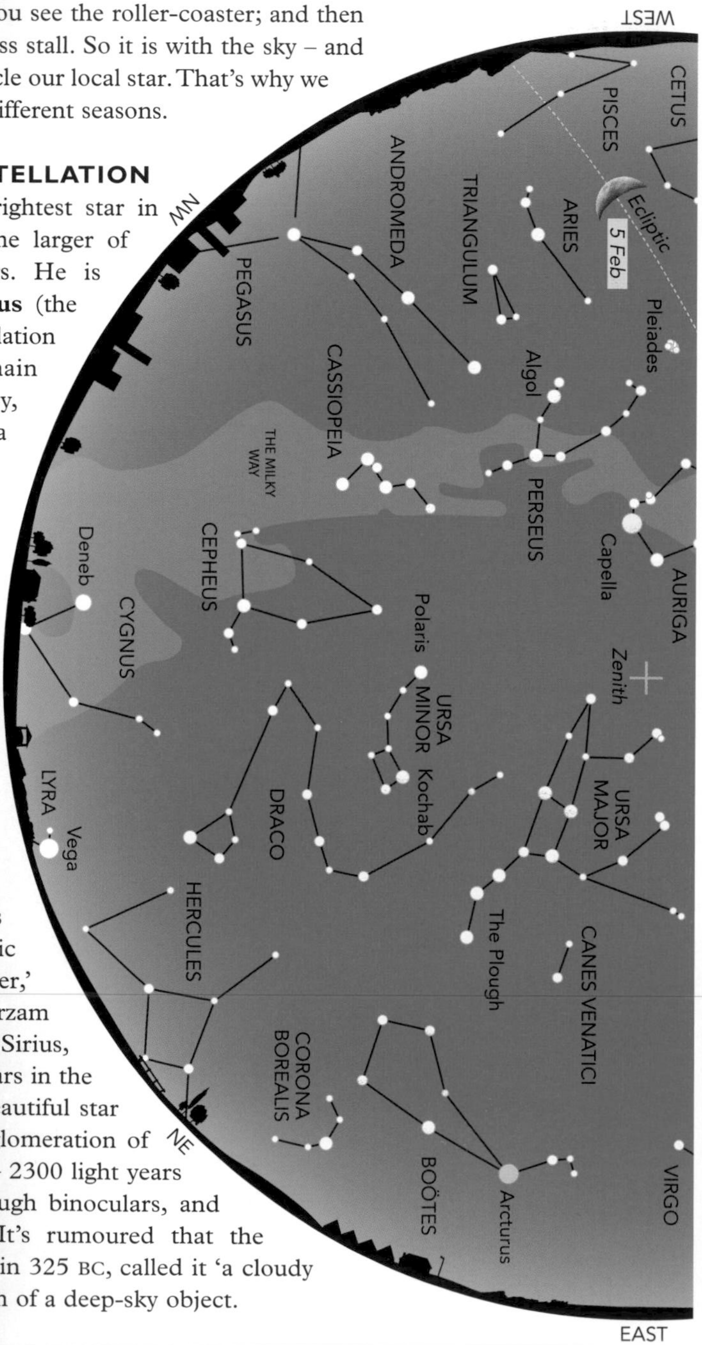

FEBRUARY'S CONSTELLATION

Crowned by **Sirius**, the brightest star in the sky, **Canis Major** is the larger of **Orion**'s two hunting dogs. He is represented as chasing **Lepus** (the Hare), a very faint constellation below Orion, but his main target is Orion's chief quarry, **Taurus** (the Bull) – take a line from Sirius through Orion's belt, and you'll spot the celestial bovine on the other side. Arabian astronomers accorded great importance to Canis Major, while the Indians regarded both cosmic dogs (**Canis Minor** lies to the left of Orion) as being 'watch-dogs of the Milky Way' – which runs between the two constellations.

To the right of Sirius is the star **Mirzam**. Its Arabic name means 'the announcer,' because the presence of Mirzam heralded the appearance of Sirius, one of the most venerated stars in the sky. Just below Sirius is a beautiful star cluster, **M41**. This loose agglomeration of over a hundred young stars – 2300 light years away – is easily visible through binoculars, and even to the unaided eye. It's rumoured that the Greek philosopher Aristotle, in 325 BC, called it 'a cloudy spot' – the earliest description of a deep-sky object.

11 pm at the beginning of February, and 9 pm at the end of the month. The planets move slightly relative to the stars during the month.

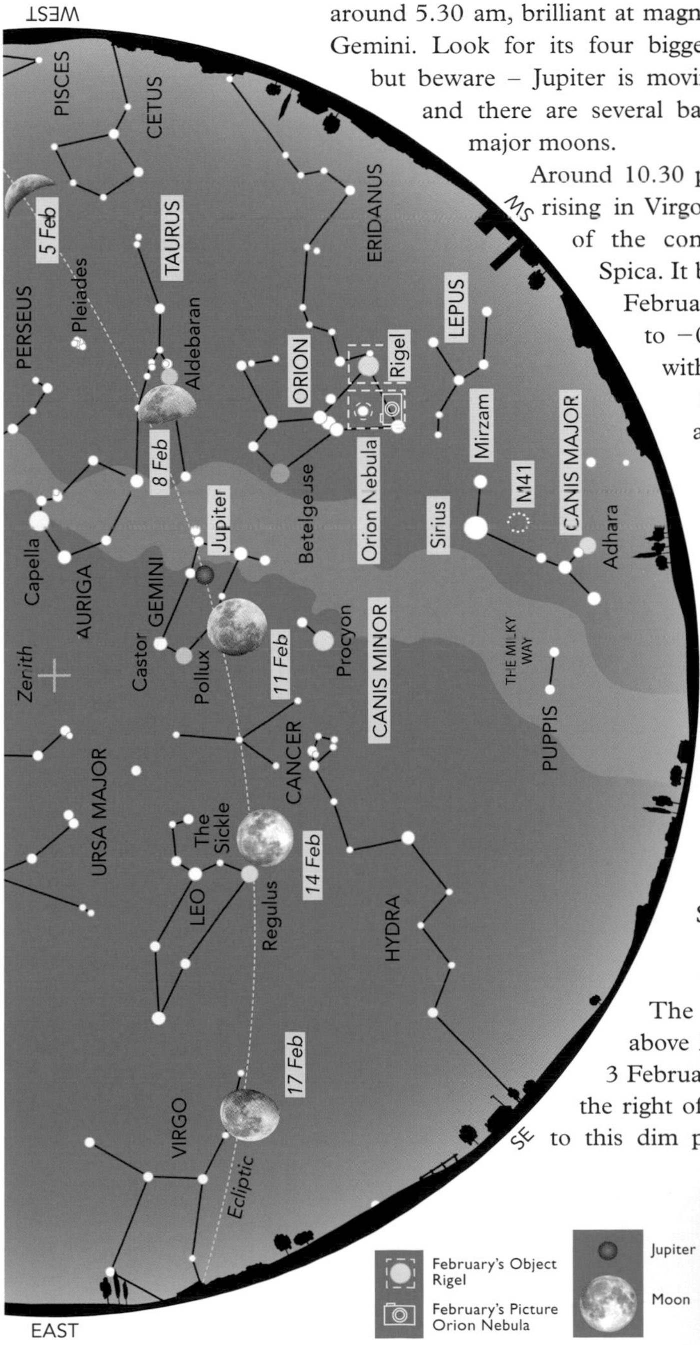

PLANETS ON VIEW

During the first week of February, seek out **Mercury**, low in the evening twilight in the west-south-west, and fading rapidly from magnitude −0.5 to +0.8.

Uranus, at magnitude +5.9 in Pisces, sets around 9.15 pm.

Jupiter hogs the limelight for most of the night, setting around 5.30 am, brilliant at magnitude −2.5 in the middle of Gemini. Look for its four biggest moons with binoculars: but beware – Jupiter is moving through the Milky Way, and there are several background stars rivalling its major moons.

Around 10.30 pm, look east to spot **Mars** rising in Virgo, a few degrees to the left of the constellation's brightest star, Spica. It brightens noticeably through February, from magnitude +0.2 to −0.5, as the Earth catches up with the Red Planet.

Saturn, in Libra, shines at magnitude +0.7 among the dull stars of Libra. The ringed planet rises at 1.00 am.

Early birds, be prepared for dazzling **Venus**, rising more than two hours before the Sun. The Morning Star blazes at magnitude −4.5, and a small telescope shows its shape change from narrow crescent to almost half-lit during February.

Neptune is lost in the Sun's glare this month.

MOON

The crescent Moon lies just above Mercury on 1 February. On 3 February, the Moon is 3 degrees to the right of Uranus – a useful signpost to this dim planet. The Moon lies near

MOON		
Date	Time	Phase
6	7.22 pm	First Quarter
14	11.53 pm	Full Moon
22	5.15 pm	Last Quarter

Aldebaran on 7 and 8 February. The brilliant object near the Moon on 10 and 11 February is Jupiter. On 14/15 February, the Moon passes below Regulus. You'll find the Moon to the right of Mars and Spica on 18 February; on 19 February, it's to the left of Spica and immediately below Mars. On the morning of 22 February, the Last Quarter Moon is close to Saturn. The crescent Moon lies near Venus before dawn on 26 February (see Special Events).

SPECIAL EVENTS

There's a spectacular pairing of Venus and the crescent Moon on the morning of **26 February**, as the two rise in the south-east around 5 am: the Morning Star is only a quarter of a degree from the horn of the narrow Moon.

FEBRUARY'S OBJECT

Rigel has to be one of the most iconic stars in the sky. Shining a dazzling blue-white, it marks the bottom right-hand corner of Orion's tunic, and it's the most brilliant star in a constellation that features a truly red-carpet list of celestial celebrities. At 860 light years distant, Rigel is about 120,000 times more luminous than our Sun. It is the seventh-brightest star in the heavens.

Rigel has been venerated in myth and legend all over the world. Its name derives from 'foot' or 'leg' in Arabic. In Central America, Rigel was known as 'the little woodpecker', while in Northern Australia, the star was named after the Red Kangaroo Leader Unumburrgu.

Rigel is actually a triple-star system. Its major companion, at magnitude +6.7, lies only 12 light days away. Although it's not particularly faint, the star is 500 times dimmer than Rigel, so you'll need a moderate telescope to spot

▼ Observing in Meifod, Wales, Peter Shah used an SXVF-H16 Starlight Xpress camera on an AG8 Orion Optics advanced astrograph to capture this image of the Orion Nebula. It was a two-panel mosaic of 6 × 10 minutes in red, 6 × 10 minutes in green, and 6 × 10 minutes in blue for each panel. The total exposure time was 6 hours and 15 minutes.

Viewing tip

Venus is a real treat this month, blazing in the morning sky. If you have a small telescope, though, don't just take a quick squint while the sky is dark: wait for the sky to brighten a bit. Seen against a black sky, the cloud-wreathed planet is so brilliant that it's difficult to make out anything on its disc. It's best to view Venus in the dawn twilight, and you can then see the planet's disc more clearly against a pale blue sky. Plus, the Morning Star will be higher in the sky and less blurred by turbulent air currents in our atmosphere.

it. Rigel B is itself double, although its companion is too close to be seen through a telescope.

Normally around 70 times bigger than the Sun, Rigel pulsates irregularly, due to disturbances in its nuclear core. As a result, its brightness changes by 0.1 of a magnitude.

FEBRUARY'S PICTURE

Before Orion sinks into the spring twilight, grab your chance to spot the iconic **Orion Nebula**. It's visible to the unaided eye below Orion's belt, but a telescope shows it at its best. This image reveals the huge gas cloud – nearly 25 light years across – with a companion nebula, NGC 1977 (top), sometimes called the Running Man Nebula. Both are essentially star-making factories, and just part of a vast region of starbirth in Orion. Young stars light up the nebulae: the dark zone between them consists of cosmic dust poised to collapse to create new stars.

FEBRUARY'S TOPIC
Aurorae

Look out for the sky's ultimate light show this year. The Aurora Borealis ('Northern Lights') and its southern equivalent – the Aurora Australis – are best seen near the Earth's northern and southern poles. In Aberdeen, these arcs, curtains and rays of green, red and yellow shifting lights are called the 'Merrie Dancers'. Aberdonian folklore puts them down to reflections of sunlight off the polar ice-cap.

The cause *is* the Sun – but in a more fundamental way. Our local star is riddled with magnetic fields, and roughly every 11 years it breaks out in a magnetic rash (see June's Object). Currently, we are near the peak of solar activity, and solar storms are frequently hurling powerful electrically-charged particles into space. If Earth is in the firing line, our planet's magnetic field channels them to the poles, where the electrical particles hit the atmosphere and light up the atoms like gas in a neon tube. Oxygen emits a green colour; nitrogen shines red.

To see the swaying celestial display is an awesome experience – and this is a great time to see the phenomenon, as sunspots are near their peak. You can take special tours by plane to the Arctic Circle – fingers crossed for clear viewing. From space, the aurorae appear as a ring of light encircling both of the Earth's polar regions.

And if you can't make it to the North Pole, never fear. Powerful aurorae have been reported from the south of France – and we've seen one in Oxfordshire!

This is *the* month to celebrate the heavens – we have a National Astronomy Week, focusing on giant planet **Jupiter**, which is at its highest in the sky for 12 years. Check out your local astronomical society, and see what star parties and presentations they have on offer. You'll have the chance to ogle the heavens though a large telescope, and get hands-on tips about observing from enthusiastic and experienced stargazers. (More details under 'Special Events'.)

▼ *The sky at 10 pm in mid-March, with Moon positions at three-day intervals either side of Full Moon. The star positions are also correct for 11 pm at*

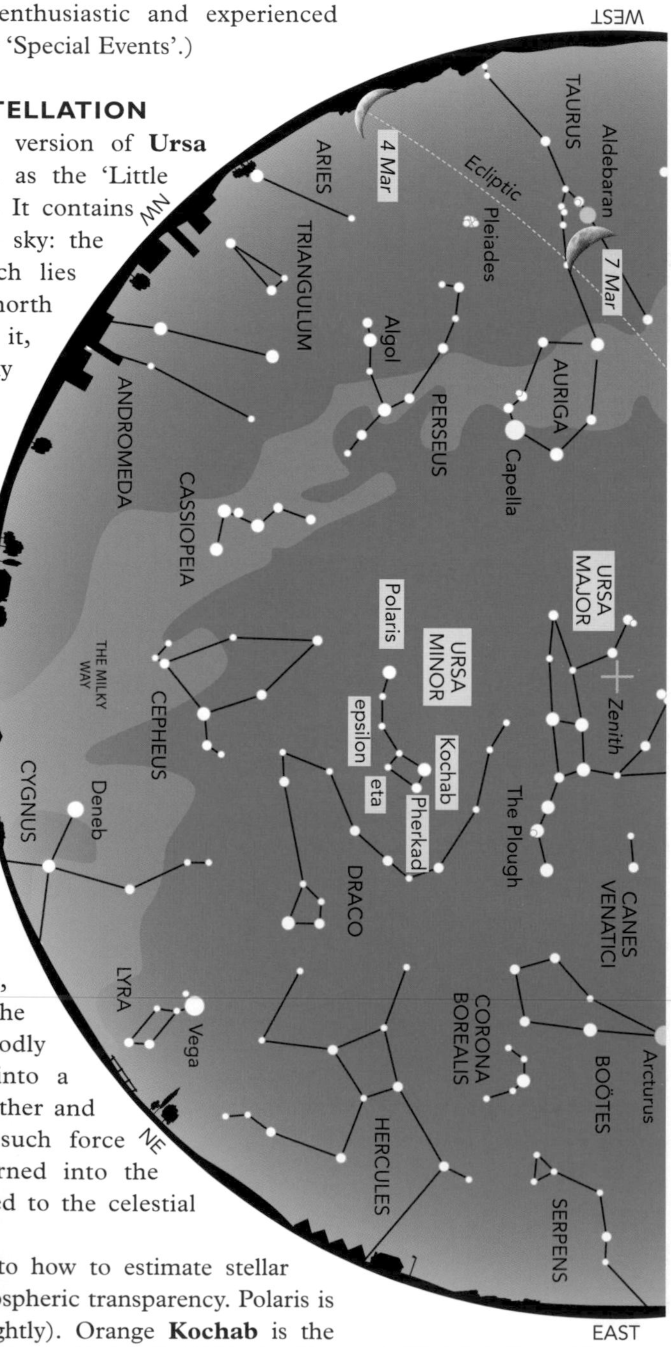

MARCH'S CONSTELLATION

Ursa Minor is a miniature version of **Ursa Major** – in fact, it's known as the 'Little Dipper' in the United States. It contains the most famous star in the sky: the Pole Star, or **Polaris**, which lies almost exactly above Earth's north pole. We spin 'underneath' it, so it is a fixed point in the sky – essential, in the past, for navigation.

In legend, Ursa Minor was Arcas – the son of the Great Bear. Originally Ursa Major (mum) was Callisto – a beautiful nymph sworn to chastity. But head god Jupiter had other ideas. Their son was born in due course – and Juno, Jupiter's wife, was so furious that she had Callisto changed into a bear.

Years later, Arcas was out hunting with his father. In the forest, he saw a bear – and began to take aim. Horrified, Jupiter realized that she was the mother of Arcas. Using his godly powers, he changed Arcas into a bear, too. Then he swung mother and son into the heavens with such force that their stumpy rumps turned into the elongated tails we see attached to the celestial bears today.

Ursa Minor is a model as to how to estimate stellar brightness – and to gauge atmospheric transparency. Polaris is magnitude +2.0 (it varies slightly). Orange **Kochab** is the

the beginning of March, and 10 pm at the end of the month (after BST begins). The planets move slightly relative to the stars during the month.

same brightness. **Pherkad** – a blue-white star – is at magnitude +3.0. **Epsilon Ursa Minoris** is a yellow giant with a brightness of magnitude +4.2. And the white star **eta** comes in last at magnitude +5.

PLANETS ON VIEW

Magnificent **Jupiter**, at magnitude −2.2 – brighter than any of the stars – is resplendent for most of the night in Gemini, only sinking below the horizon at 3.30 am (see Topic).

But it's now being challenged by **Mars,** heading for opposition next month. The Red Planet rises around 10 pm at the start of March, but as early as 7.30 pm by month's end. As it steams through Virgo – just a few degrees from Spica – its brightness soars from magnitude −0.5 to −1.3.

Saturn (magnitude +0.6) lies in Libra, rising about 11 pm.

Brilliant **Venus** rises as the Morning Star around 4.30 am. It shines at magnitude −4.3, and reaches greatest western elongation on 22 March.

Uranus, at magnitude +5.9 in Pisces, sets at 8.30 pm at the beginning of March, and quickly disappears into the twilight glow.

Neptune is invisible in the Sun's glare in March; as is **Mercury**, even though the innermost planet is at greatest western elongation on 14 March.

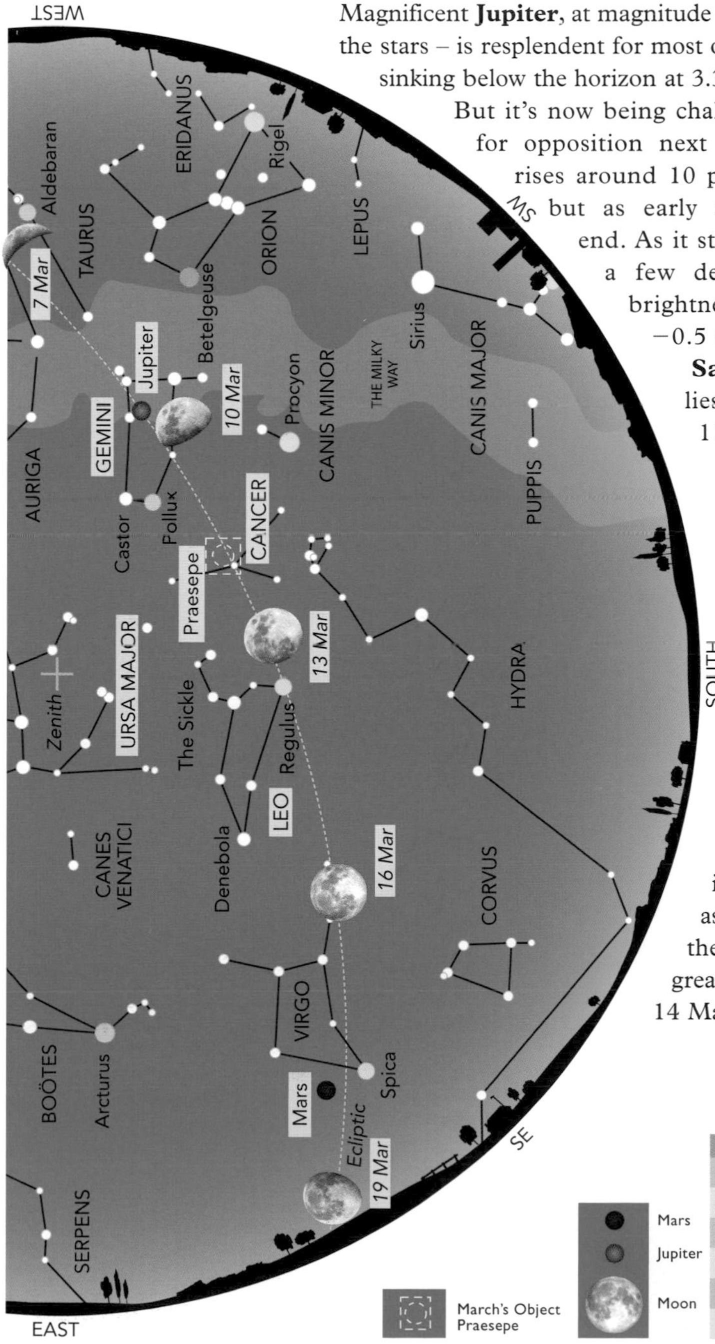

MOON		
Date	Time	Phase
1	8.00 am	New Moon
8	1.27 pm	First Quarter
16	5.08 pm	Full Moon
24	1.46 am	Last Quarter
30	7.44 pm	New Moon

MOON

On 7 March, the Moon passes close to Aldebaran, through the fringes of the Hyades star cluster. It lies near Jupiter on 9 and 10 March. On 14 March, the Moon moves below Regulus. It lies close to Spica on 18 March, with Mars to the upper left. On the night of 20/21 March, the Moon nuzzles up close to Saturn. The crescent Moon lies just above Venus on the morning of 27 March.

SPECIAL EVENTS

1–8 March is National Astronomy Week, when astronomical societies and observatories – great and small – will be welcoming visitors to explore the heavens. More details can be found at www.astronomyweek.org.uk.

The Vernal Equinox, on **20 March** at 4.57 pm, marks the beginning of spring, as the Sun moves up to shine over the northern hemisphere.

30 March, 1.00 am: British Summer Time starts – don't forget to put your clocks forward (the mnemonic is 'Spring forward, Fall back').

MARCH'S OBJECT

Between **Gemini** and **Leo** lies the faint zodiacal constellation of **Cancer** (the Crab). You'd be hard pressed to see it from a city, but try to concentrate your eyes on the central triangle of stars – then look inside. With the unaided eye, you can see a misty patch. This is **Praesepe** – a dense group of stars whose name literally means 'the manger', but is better known as the Beehive Cluster. If you train binoculars on it, you'll understand how it got its name – it really does look like a swarm of bees.

Praesepe lies nearly 600 light years away, and contains over 1000 stars, all of which were born together some 600 million years ago. Two of its stars have planets in orbit about them but they are not 'Earths' – instead, they are 'hot Jupiters', gas giants circling close in to their parent star.

Galileo, in 1610, was the first to recognize Praesepe as a star cluster. But the

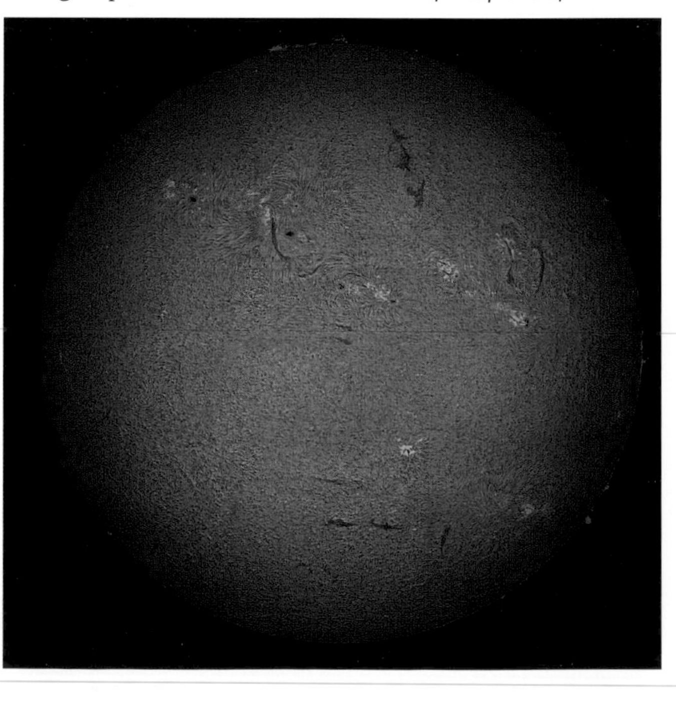

▼ Dave Tyler of High Wycombe, Buckinghamshire, took this sensational image of the Sun, in the light of hydrogen alpha, on 2 April 2013. He used a Coronado 90 telescope coupled to a Point Grey Flea3 webcam. The image is a stack of the best 900 individual shots out of a sequence of 3000.

> 👁 ***Viewing tip***
> This is the time of year to tie down your compass points – the directions of north, south, east and west as seen from your observing site. North is easy – just latch on to Polaris, the Pole Star. And at noon, the Sun is always in the south. But the useful extra in March is that we hit the Spring (Vernal) Equinox, when the Sun rises due east, and sets due west. So remember those positions relative to a tree or house around your horizon.

ancient Chinese astronomers obviously knew about it, naming the cluster Zei She Ge – 'the Exhalation of Piled-up Corpses'!

MARCH'S PICTURE

Our Sun is making a determined appearance in the skies at last – and, currently, it's quite an active star (see June's Object). Magnetic forces dam back its energy to create dark sunspots. This image, in a wavelength emitted by hydrogen atoms, reveals bright regions above the sunspots, along with giant loops of gas – seen as dark filaments against the Sun, and glowing prominences at the rim – which show how its magnetic power extends far out into space.

MARCH'S TOPIC
Jupiter

Jupiter is riding higher in the sky than at any time in the past decade – and that's why it's the focus of this year's National Astronomy Week. The planet is hardly our closest neighbour – it lies about 700 million kilometres away – but it is vast. At 143,000 kilometres in diameter, it could contain 1300 Earths – and as it's made almost entirely of gas, it's very efficient at reflecting sunlight. So it's a fantastic target for stargazers, whether you are using your unaided eyes, binoculars or a small telescope.

Although Jupiter is so huge, it spins faster than any other planet in the Solar System. It rotates every 9 hours 55 minutes, and as a result its equator bulges outwards – through a small telescope, it looks a bit like a tangerine crossed with an old-fashioned humbug. The humbug stripes are cloud belts of ammonia and methane stretched out by the planet's dizzy spin.

Jupiter has a fearsome magnetic field that no astronaut would survive, huge lightning storms, and an internal heat source which means it radiates more energy than it receives from the Sun. Jupiter's core simmers at a temperature of 20,000°C.

Jupiter commands its own 'mini solar system' – a family of almost 70 moons. The four biggest are visible in good binoculars, and even – to the really sharp-sighted – to the unaided eye. These are worlds in their own right (Ganymede is even bigger than the planet Mercury). But two vie for 'star' status: Io and Europa. The surface of Io is erupting, with incredible geysers ejecting plumes of sulphur dioxide 300 kilometres into space. Brilliant white Europa probably contains oceans of liquid water beneath a solid ice coating, where alien fish may swim....

After all the attention focused on giant gas-world **Jupiter** in the last few months, it's the time for **Mars** to shine. In early April, the Red Planet is closer to the Earth than it has been for six years.

Mars is currently the jewel adorning the ancient constellation **Virgo** (the Virgin), which dominates the springtime skies along with her companion, **Leo**. While Leo does indeed look like a recumbent lion, it's hard to envisage Virgo as anything other than a vast 'Y' in the sky!

▼ The sky at 11 pm in mid-April, with Moon positions at three-day intervals either side of Full Moon. The star positions are also correct for midnight at the beginning of

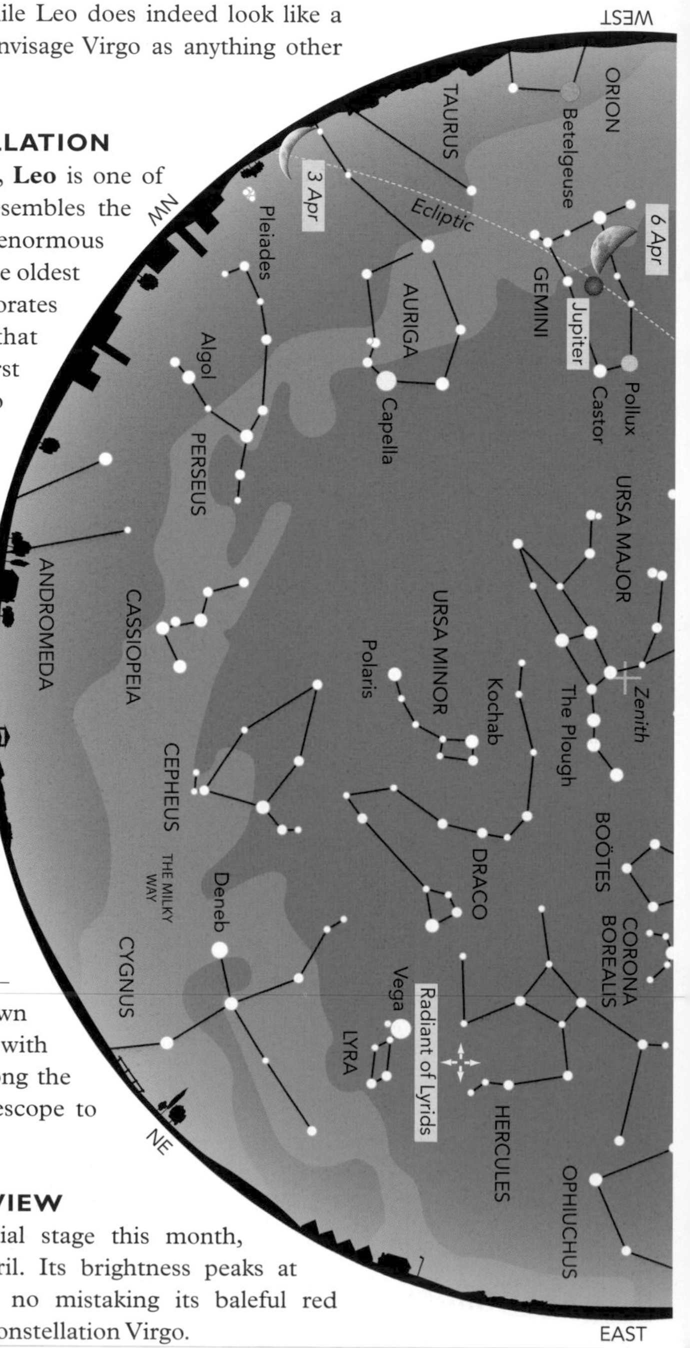

APRIL'S CONSTELLATION

Like the mighty hunter Orion, **Leo** is one of the rare constellations that resembles the real thing – in this case, an enormous crouching lion. Leo is among the oldest constellations, and commemorates the giant Nemean lion that Hercules slaughtered as the first of his 12 labours. According to legend, the lion's flesh could not be pierced by iron, stone or bronze – so Hercules wrestled with the lion and choked it to death.

The lion's heart is marked by the first-magnitude star **Regulus**, and its other end by **Denebola**, which means in Arabic 'the lion's tail'. A small telescope shows that **Algieba**, the star marking the lion's shoulder, is actually a beautiful close double star. Just underneath the main 'body' of Leo are several spiral galaxies – nearby cities of stars like our own Milky Way. They can't be seen with the unaided eye, but sweep along the lion's tummy with a small telescope to reveal them.

PLANETS ON VIEW

Mars roars on to the celestial stage this month, reaching opposition on 8 April. Its brightness peaks at magnitude −1.5, and there's no mistaking its baleful red presence all night long in the constellation Virgo.

April, and 10 pm at the end of the month. The planets move slightly relative to the stars during the month.

As Mars rises earlier each evening, **Jupiter** – in Gemini – sinks lower in the west, setting around 2.30 am. Though it's now on the way out, the giant planet is still a tad brighter than the Red Planet, at magnitude −2.0.

Saturn (magnitude +0.4) lies in Libra, and rises around 10 pm.

Glorious **Venus** – rising around 5 am – is slipping down into the morning twilight, where it will hang around for the next six months. The Morning Star shines at magnitude −4.1.

Mercury, **Uranus** and **Neptune** are lost in the Sun's glare in April.

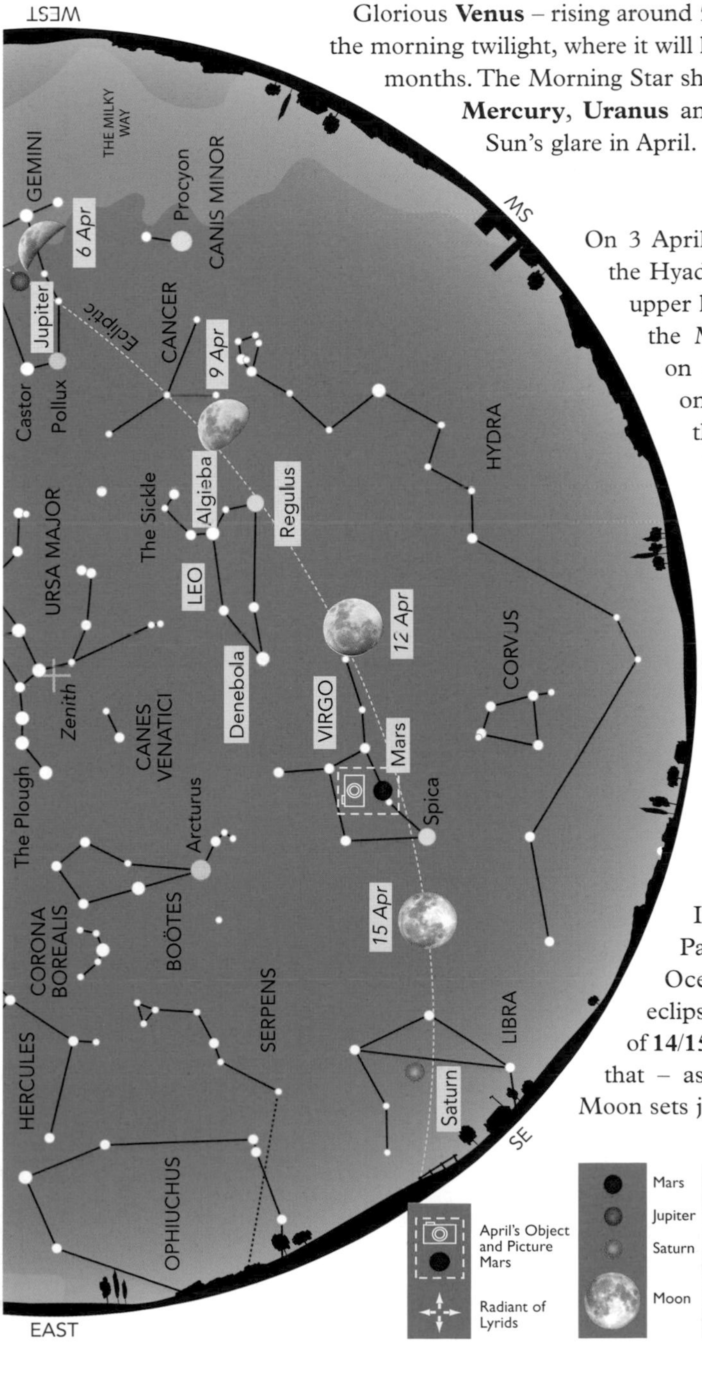

MOON

On 3 April, the Moon encroaches on the Hyades; on 4 April, it lies to the upper left of Aldebaran. You'll find the Moon right next to Jupiter on 6 April. It is near Regulus on 10 April. On 14 April, the almost-Full Moon lies between Mars (upper right) and Spica (lower left), skimming past the latter in the early hours of 15 April. The Moon is near Saturn on 16 April. The reddish star near the Moon on the night of 18/19 April is Antares. The crescent Moon lies near Venus before dawn on 25 and 26 April.

SPECIAL EVENTS

If you're in the Americas, the Pacific or the western Atlantic Ocean, look out for a total eclipse of the Moon on the night of **14/15 April**. We're sorry to report that – as seen from the UK – the Moon sets just after the eclipse begins.

MOON		
Date	Time	Phase
7	9.31 am	First Quarter
15	8.42 am	Full Moon
22	8.51 am	Last Quarter
29	7.14 am	New Moon

▲ *Using an Atik webcam to provide video sequences for each individual shot, Dave Tyler captured these images of Mars using a 355 mm Schmidt-Cassegrain telescope sited in High Wycombe, Buckinghamshire. This is an inverted view through an astronomical telescope, with north at the bottom.*

21/22 April: It's the maximum of the **Lyrid** meteor shower, which – by perspective – appears to emanate from the constellation of Lyra. The shooting stars are dust particles from Comet Thatcher. Best viewed late in the evening, before the Moon rises; expect some meteors tomorrow night as well.

There's an eclipse of the Sun on **29 April** – but don't hold your breath! It's an annular eclipse (where a ring of the Sun's surface is visible around the Moon) and you'll only get to see that from Antarctica. People in Australia and the south Atlantic and Indian Oceans will experience a partial eclipse – nothing is visible from the UK.

APRIL'S OBJECT

On 8 April, **Mars** is at opposition – in line with the Sun and Earth – though its elliptical orbit means the Red Planet is actually closest to Earth (92 million kilometres away) six days later. Use a small telescope to skim over its mottled surface, and spot its icy polar caps.

The debate about ice, water and life on Mars has hotted up over the last few years. There's evidence from NASA's Viking landings in 1976 that primitive bacterial life was then still in existence. And the present flotilla of space probes, which are crawling over its surface or orbiting the Red Planet, are

Viewing tip
When you first go out to observe, you may be disappointed at how few stars you can see in the sky. But wait for around 20 minutes and you'll be amazed at how your night vision improves. One reason for this 'dark adaption' is that the pupil of your eye gets larger to make the best of the darkness. More importantly, in dark conditions the retina of your eye builds up much bigger reserves of rhodopsin, the chemical that responds to light.

unanimously picking up evidence for present or past water all over Mars – the essential ingredient for life.

NASA's Curiosity mission, which landed on Mars in August 2012, is actively sniffing around Gale Crater. Its rover – the size of a Mini Cooper car – is investigating the geology, composition and life-potential of the Red Planet. And for the first time, NASA admits it has designed the project with a view that looks at the possibility of humans going to Mars.

APRIL'S PICTURE

'Star' of the month: **Mars**, the Red Planet. This sequence of images, taken when Mars was close to Earth in 2005–6, shows how the planet's apparent size changes as the two worlds move with respect to each other. The markings on Mars – which were once thought to be vegetation – are now known to be expanses of dark rocks. Mars has white polar caps made of frozen carbon dioxide and ice – which show up particularly well in the far right-hand image. And the atmospheric haze looks spectacular around Mars's lower pole.

APRIL'S TOPIC
Star colours

Spring is well and truly here, with Arcturus gracing our night skies again. You can't miss this brilliant beacon high in the south-east – but look carefully at its colour. This star isn't plain, boring white. Arcturus is distinctly orange-red: a sure sign that it's a cool star. Its surface temperature is around 4000°C, as compared to 5500°C for our yellow Sun.

Star colours are a good guide to their temperatures. The hottest stars are blue-white. White stars come next; then yellow, orange and red.

Arcturus is a red giant: a distended star close to the end of its life. Its atmosphere has expanded, and cooled. But it's no match for the biggest of the red giants, where the temperature of their bloated outer layers has fallen to little more than 3000°C – and as a result they shine blood-red. Take a look at Betelgeuse in Orion (see November's Object), now just setting in the west; or Antares in Scorpius, rising in the south-east about 1 am, which takes its very name from its baleful-red colour – 'Anti-Ares', the 'rival of Mars'. If either of these stars were placed in the Solar System, it would stretch all the way to the asteroid belt!

Contrast these stars with Spica, the bright, conspicuously blue-white star at the heart of Virgo. This is a star in the prime of life. Spica boasts a surface temperature of 22,000°C, and is more than 12,000 times brighter than the Sun.

Ringworld **Saturn** is at its closest this month, heading up a display of all five naked-eye planets. You can catch Mercury and Jupiter at dusk; Mars in the evening sky with Saturn; and Venus rising just before the Sun. And we *may* be in for an unexpected display of celestial fireworks near the end of May.

A sure sign that warmer days are here is the appearance of **Arcturus** – a distinctly orange-coloured star that lords it over a huge area of sky devoid of other bright stars. Summer is on the way!

▼ *The sky at 11 pm in mid-May, with Moon positions at three-day intervals either side of Full Moon. The star positions are also correct for midnight at the beginning of*

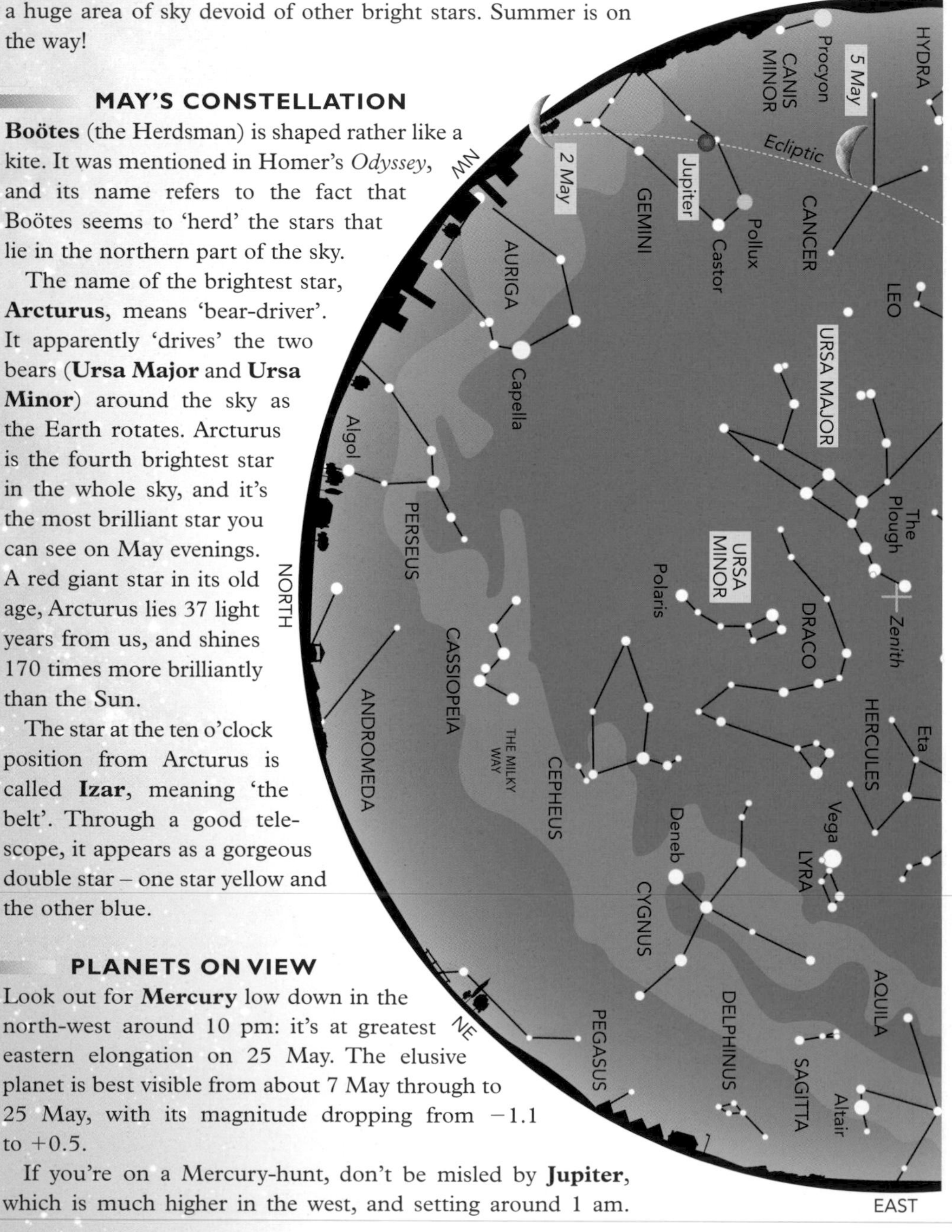

MAY'S CONSTELLATION

Boötes (the Herdsman) is shaped rather like a kite. It was mentioned in Homer's *Odyssey*, and its name refers to the fact that Boötes seems to 'herd' the stars that lie in the northern part of the sky.

The name of the brightest star, **Arcturus**, means 'bear-driver'. It apparently 'drives' the two bears (**Ursa Major** and **Ursa Minor**) around the sky as the Earth rotates. Arcturus is the fourth brightest star in the whole sky, and it's the most brilliant star you can see on May evenings. A red giant star in its old age, Arcturus lies 37 light years from us, and shines 170 times more brilliantly than the Sun.

The star at the ten o'clock position from Arcturus is called **Izar**, meaning 'the belt'. Through a good telescope, it appears as a gorgeous double star – one star yellow and the other blue.

PLANETS ON VIEW

Look out for **Mercury** low down in the north-west around 10 pm: it's at greatest eastern elongation on 25 May. The elusive planet is best visible from about 7 May through to 25 May, with its magnitude dropping from −1.1 to +0.5.

If you're on a Mercury-hunt, don't be misled by **Jupiter**, which is much higher in the west, and setting around 1 am.

May, and 10 pm at the end of the month. The planets move slightly relative to the stars during the month.

The giant planet shines at magnitude −1.8, among the stars of Gemini.

Over the other side of the sky, **Mars** – after its close approach in April – starts the month almost equally bright, at magnitude −1.2. But its brightness quickly falls away, to end May at magnitude −0.5. Moving through Virgo, the Red Planet sets around 5 am.

With all this competition from its brighter sibling planets, poor **Saturn** is a tad overshadowed even though it is at its closest and brightest this month. The ringworld, in Libra, is at opposition on 10 May, shines at magnitude +0.2 and is visible all night long.

Venus, at magnitude −3.9, is very low in the dawn twilight, rising just an hour before the Sun.

Uranus and **Neptune** are too close to the Sun to be seen this month.

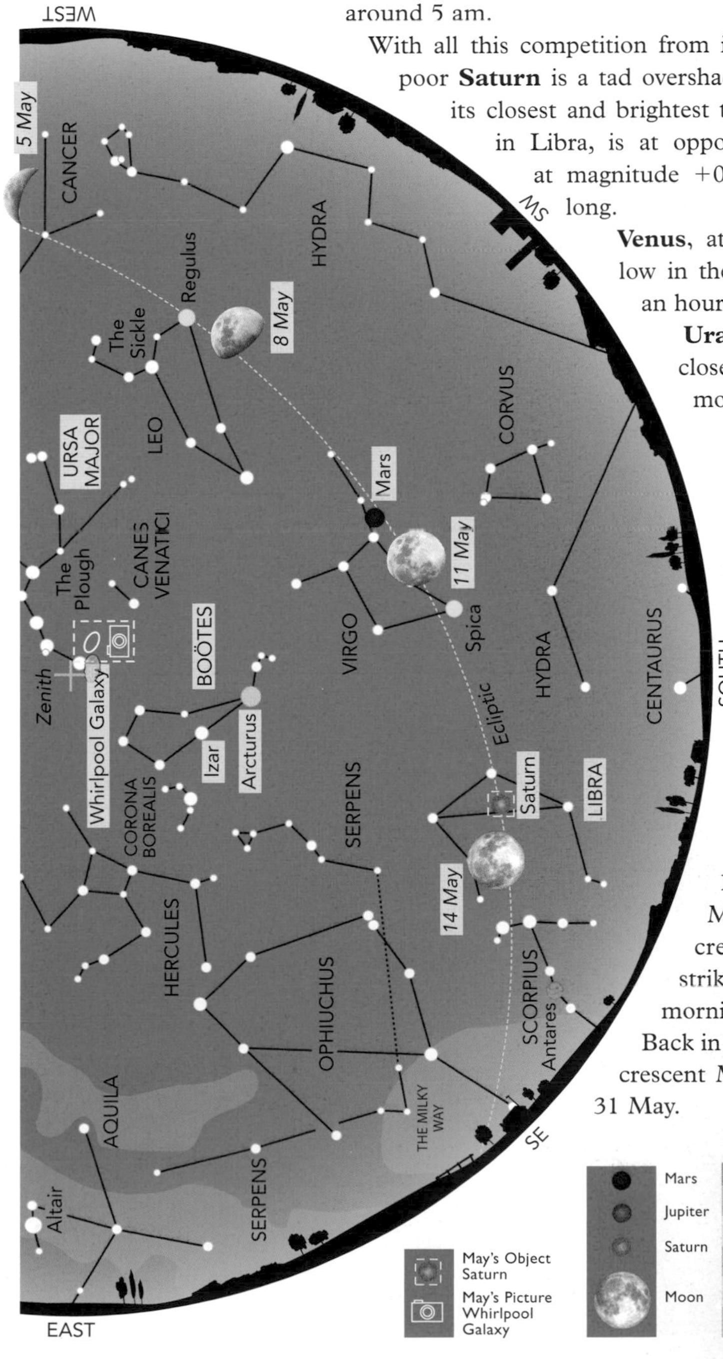

MOON

The crescent Moon lies near Jupiter on 3 and 4 May. On 7 and 8 May, it passes below Regulus. You'll find the Moon to the right of Mars on 10 May; on 11 May, it's between Mars and Spica; and on 12 May, the Moon is just to the left of Spica. Saturn is near the Moon on 13 and 14 May. On 15 May, the Moon is above Antares. The crescent Moon makes a striking pair with Venus in the morning sky of 25 and 26 May. Back in the evening sky, the waxing crescent Moon lies below Jupiter on 31 May.

MOON		
Date	Time	Phase
7	4.15 am	First Quarter
14	8.16 pm	Full Moon
21	1.59 pm	Last Quarter
28	7.40 pm	New Moon

SPECIAL EVENTS

The maximum of the Eta Aquarid meteor shower falls on **5/6 May**. These shooting stars – tiny pieces shed by Halley's Comet and burning up in Earth's atmosphere – are best seen after 2 am, when the Moon has set.

Around **22 May**, the European Space Agency's deep-space mission Rosetta arrives at Comet Churyumov–Gerasimenko, becoming the first spacecraft to orbit a comet's nucleus. It will observe the burgeoning activity of this 'dirty snowball' on its way to closest approach to the Sun in August 2015.

And – look out on the night of **23/24 May** for a possible storm of meteors! The Earth will run into debris shed by an obscure comet that was discovered in 2004 by an automated sky survey, Comet LINEAR. Predictions are quite uncertain, but after midnight we *may* see over 1000 meteors falling per hour.

Viewing tip

Have a meteor party to check out the possible shooting-star storm on 23/24 May! You don't need any optical equipment – in fact, telescopes and binoculars will restrict your view of the meteor shower. The ideal viewing equipment is your unaided eye, plus a sleeping bag and a lounger on the lawn. If you want to make measurements, a stopwatch and clock are good for timings, while a piece of string will help to measure the length of the meteor trails.

MAY'S OBJECT

The slowly-moving ringworld **Saturn** is currently livening up the dull constellation of **Libra** (the Scales). It's famed for its huge engirdling appendages: the rings would stretch nearly all the way from the Earth to the Moon. The planet is a glorious sight through a small telescope, like an exquisite model hanging in space.

And the rings are just the beginnings of Saturn's larger family. It has at least 62 moons, including Titan – which is also visible through a small telescope. The international Cassini–Huygens mission has discovered lakes of liquid methane and ethane on Titan, and possibly active volcanoes.

And the latest exciting news is that Cassini has imaged plumes of salty water spewing from its icy moon Enceladus, while Dione has traces of oxygen in its thin atmosphere. These discoveries raise the intriguing possibility of primitive life on Saturn's moons.

Saturn itself is second only to Jupiter in size. But it's so low in density that were you to plop it in an ocean, it would float. Like Jupiter, Saturn has a ferocious spin rate – 10 hours and 32 minutes – and its winds roar at speeds of up to 1800 km/h.

Saturn's atmosphere is much blander than that of its larger cousin. But it's wracked with lightning-bolts 1000 times more powerful than those on Earth.

MAY'S PICTURE

Just underneath the 'tail' of the Great Bear lies this celestial gem – the **Whirlpool Galaxy**, in Canes Venatici (the Hunting Dogs). Visible through binoculars, this glorious celestial Catherine wheel – often known by its catalogue number, M51 – lies over 20 million light years away. Its phenomenal spiral arms are a result of the gravitational tug of a smaller companion galaxy, NGC 5195, which has just whizzed past the Whirlpool.

◀ *Nick Hart captured this image of the famous Whirlpool Galaxy. He used a 250 mm f/4.8 reflecting telescope equipped with an Atik 314L mono CCD camera, plus red, green and blue filters. This image was taken from a light-polluted location (Newport, in South Wales), but the photographer has used his skill to bring out the galaxy's delicate tracery.*

MAY'S TOPIC
The Rosetta mission

This month, Europe's robotic probe Rosetta swings into orbit around a comet with a memorable title – Comet Churyumov–Gerasimenko – named after its discoverers at the Almaty Observatory in Kazakhstan. Understandably, it's been nicknamed 'Comet C–G' for the mission!

Comets – dirty balls of ice and rock – are the building blocks of our Solar System. More than this, they may well contain the building blocks of life. Comets almost certainly delivered the waters that clothe our Earth in its oceans, along with the carbon-rich compounds that make up living cells.

Rosetta's mission – hence its name – is to look for origins. Just as the Egyptian Rosetta Stone unlocked the early history of human civilization, the Rosetta probe is investigating the beginnings of life in space.

And there's more to come. Later this year (see November's Special Events), Rosetta will gently land a small spacecraft on the comet, which is only 4 kilometres wide. The lander, Philae – named after the island on the Nile where archaeologists discovered an obelisk that enabled them to decode the Rosetta Stone – will study the structure, physics and chemistry of Comet C–G in depth.

This month isn't the best time of year to go stargazing. The Sun reaches its highest position over the northern hemisphere in June, so we get the longest days and the shortest nights.

But take advantage of the soft, warm weather to acquaint yourself with the lovely summer constellations of **Hercules**, **Scorpius**, **Lyra**, **Cygnus** and **Aquila** – along with the bright planets **Mars** and **Saturn**.

▼ *The sky at 11 pm in mid-June, with Moon positions at three-day intervals either side of Full Moon. The star positions are also correct for midnight at the beginning of*

JUNE'S CONSTELLATION

For one of antiquity's superheroes, the celestial version of **Hercules** looks like a wimp. While Orion is all strutting masculinity, Hercules is but a poor reflection – and upside-down to boot.

The two constellations are similar in shape – you can see the outline of a man up there – but the stars are faint and undistinguished. Shame: because Hercules was one of the ancient Greeks' main legends, famous for his 12 labours of heroism.

However, dig a little deeper, and you'll find a fascinating constellation. Outside the rectangular main 'body' of the hero, and to the south, you'll find **Rasalgethi** – Hercules' head. At around 400 times the Sun's girth, it is one of the biggest stars known. This distended object, close to the end of its life, flops and billows in its death throes. As a result, Rasalgethi varies in brightness, changing from third to fourth magnitude over a period of about 90 days.

Hercules boasts one of the most spectacular sights in the northern night sky. Go back to 'the rectangle' and look about a quarter of the way down from the top right-hand star (**eta Herculis** – a sun-like star), and you'll see a fuzzy patch. **M13** is a globular cluster made of almost a million stars. These are some of the oldest stars in our Galaxy –

June, and 10 pm at the end of the month. The planets move slightly relative to the stars during the month.

a bee-like swarm of red giants. It's really great to gaze at with a small telescope.

PLANETS ON VIEW

After months of splendiferous glory in Gemini, **Jupiter** is now sinking down to rest. At the beginning of June, the giant planet (magnitude −1.7) is setting at midnight; by the end of the month, it has disappeared from view in the twilight glow.

Mars is now king of the evening sky, strutting its stuff in Virgo and setting around 1.30 am. During June, the Red Planet fades from magnitude −0.5 to 0.0.

Ringworld **Saturn** lies in Libra. At magnitude +0.4, it sets around 3 am.

Venus rises just after 3 am; though it's low in the dawn twilight, the Morning Star (magnitude −3.8) is bright enough to be easily visible if you have a clear horizon.

Mercury, **Uranus** and **Neptune** are lost in the Sun's glare throughout June.

MOON

On 1 June, the crescent Moon lies to the left of Jupiter. The Moon is near Regulus on 4 June. You'll find it right below Mars on 7 June, and very close to Spica on 8 June. The Moon is near Saturn on 10 June. On 11 and 12 June, it passes above Antares. On the night of 20/21 June, use the Moon to locate Uranus, about a Moon's-width below our celestial companion (scan with binoculars). The crescent Moon lies near Venus on the morning of 24 June.

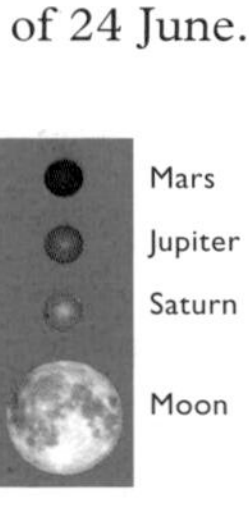

MOON		
Date	Time	Phase
5	9.39 pm	First Quarter
13	5.11 am	Full Moon
19	7.39 pm	Last Quarter
27	9.08 am	New Moon

▲ *Image of noctilucent clouds captured over St Lawrence Bay in Essex, on 13 June 2004. Robin Scagell was using a Canon 10D camera at ISO 1600 with a 15-second exposure at f/4.5.*

SPECIAL EVENTS

21 June, 11.51 am: Summer Solstice. The Sun reaches its most northerly point in the sky, so 21 June is Midsummer's Day, with the longest period of daylight. Correspondingly, we have the shortest nights.

JUNE'S OBJECT

The **Sun** is livening up – and in more ways than one. At the height of summer, our local star rides high in the sky, and we feel the heat of its rays. Some 150 million kilometres away, the Sun is our local star – and our local nuclear reactor.

At its core, where temperatures reach 15.7 million degrees, this giant ball of gas fuses atoms of hydrogen into helium. Every second, it devours 4 million tonnes of itself, bathing the Solar System with light and warmth.

But the Sun is also a dangerous place. It's now near the peak of its activity, in a cycle that repeats roughly every 11 years. The driver is the Sun's magnetic field, wound up by the spinning of our star's surface gases. The magnetic activity suppresses the Sun's circulation, leading to a rash of dark sunspots. Then the pent-up energy is released in a frenzy of activity, when our star hurls charged particles through the Solar System. These dangerous particles can kill satellites, and disrupt power lines on Earth. And if we are ever to make the three-year human journey to Mars and back, we will have to take the Sun's unpredictable, malevolent, weather into account.

And the Sun's extreme heat makes it hazardous to observe without special precautions – follow the advice in this month's Viewing Tip for a safe view of our local star.

Viewing tip

June is *the* month for the best Sun-viewing, and this year is a great time to catch up with our local star, as its sunspot activity (and associated flares and eruptions) is near its (roughly) 11-year maximum. But be careful. **NEVER** use a telescope or binoculars to look at the Sun directly: it could blind you permanently. Fogged film is no safer, because it allows the Sun's infrared (heat) rays to get through. Eclipse goggles are safe (unless they're scratched). The best way to observe the Sun is to project its image through binoculars or a telescope on to a white piece of card. Or – if you want the real 'biz' – get some solar binoculars or even a solar telescope, with filters that guarantee a safe view. Check the web for details.

JUNE'S PICTURE

This is the month to look out for **noctilucent clouds**. These blue, eerie-looking sky-sights can be seen in the late evening twilight, after all the other clouds have disappeared. June is the best time to catch them, when the Sun isn't too far below the horizon. These are the highest clouds in Earth's atmosphere, shining by reflected sunlight at an altitude of around 80 kilometres. There's still no agreement about their origin. They seem to be composed of ice, frozen on to minute dust particles. But are they from volcanic eruptions – or from micro-meteoroids?

JUNE'S TOPIC
Search for Extra-Terrestrial Intelligence

Whenever we give a presentation, one of the questions we're most asked is: 'Is there anybody out there?' It's now half a century since the veteran American astronomer Frank Drake turned his radio telescope to the heavens in the hope of hearing an alien broadcast. Despite false alarms (probably caused by secret military equipment), there has been a deafening silence.

Drake and his colleagues founded an independent institution in California – the SETI Institute (Search for Extra-Terrestrial Intelligence). It's a serious scientific endeavour that looks at the biology, psychology and motivation into our search for alien life.

Recently, their fortunes have been boosted by a donation from Paul Allen (who co-founded Microsoft). Thanks to Allen, the team is now building an array of 400 radio telescopes in California to tune into that first whisper from ET.

But is life out there far more advanced than us? Is radio communication something that came and went? The SETI researchers are contemplating communication by laser beams – but even that may prove too primitive.

High summer is here, and with it comes the brilliant trio of the **Summer Triangle** – the stars **Vega**, **Deneb** and **Altair**. Each is the brightest star in its own constellation: Vega in **Lyra**, Deneb in **Cygnus**, and Altair in **Aquila**. And this is the time to catch the far-southern constellations of **Sagittarius** and **Scorpius** – embedded in the glorious heart of the Milky Way.

▼ *The sky at 11 pm in mid-July, with Moon positions at three-day intervals either side of Full Moon. The star positions are also correct for midnight at the beginning of*

JULY'S CONSTELLATION

Ursa Major (the Great Bear) is one of everyone's favourite star patterns. Its seven brightest stars are usually called **'the Plough'**. But – unlike the brilliant stars of Orion, which make up the shape of a convincing superhero – those of the Plough are fainter. And most people today have probably never seen an old-fashioned, horse-drawn plough, from which the constellation takes its name. In fact, some children call it 'the saucepan', while in the United States it's known as 'the Big Dipper'.

But the Plough is the first constellation that most people get to know. There are two reasons. Firstly, it's always on view in the northern hemisphere. And secondly, the two end stars of the 'bowl' of the Plough point directly towards the Pole Star, **Polaris**.

Though it seems so familiar, Ursa Major is unusual in a couple of ways. It contains a double star that you can actually split with the naked eye. **Mizar**, the star in the middle of the bear's tail (or the handle of the saucepan), has a fainter companion, **Alcor**. The whole system – once thought to be a chance alignment – consists of six stars.

In addition, unlike most of the constellations, the majority of the stars in the Plough lie at the same distance and were born together. Leaving aside the two end stars, **Dubhe** and **Alkaid**, the others are all moving in the same direction (along with brilliant Sirius, which is also a member of

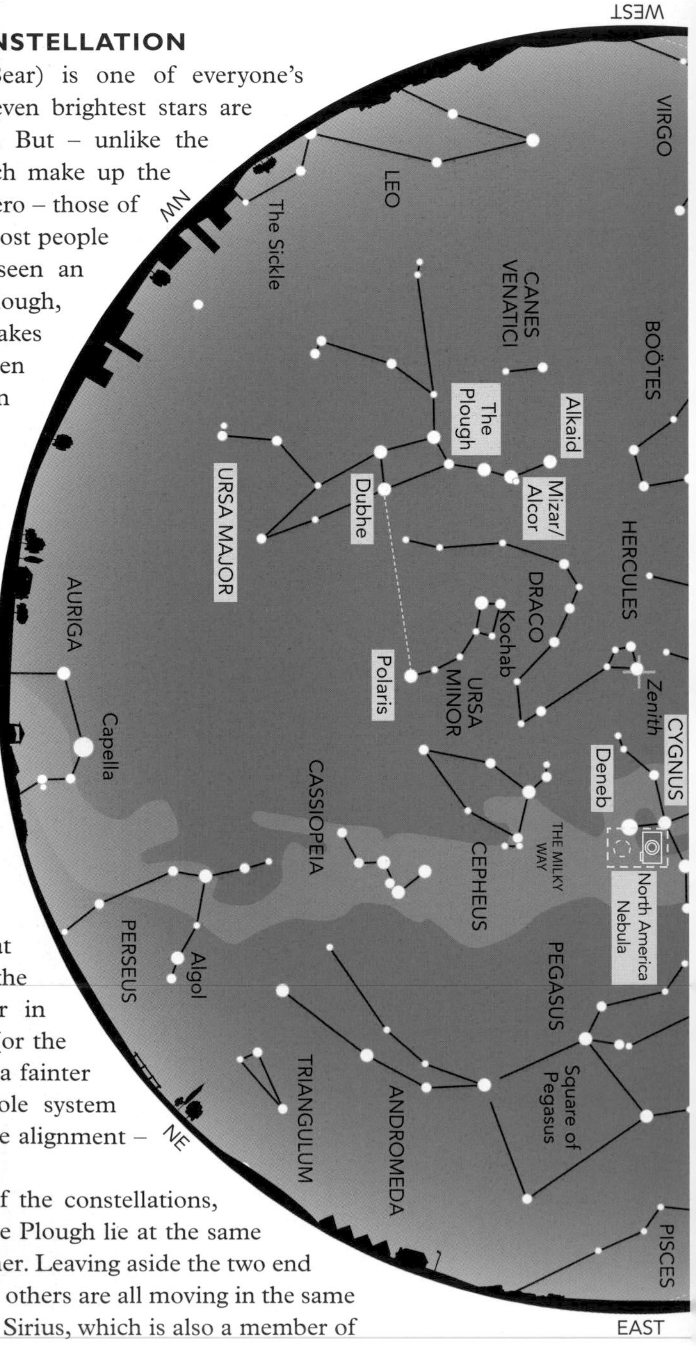

July, and 10 pm at the end of the month. The planets move slightly relative to the stars during the month.

the group). Over thousands of years, the shape of the Plough will gradually change, as Dubhe and Alkaid go off on their own paths.

PLANETS ON VIEW

The bright reddish 'star' to the south-west in the evening is the planet **Mars**, now setting around midnight. As the Earth draws away from the slow-moving planet, its brightness drops from magnitude 0.0 to +0.4. Lying in Virgo, the Red Planet passes bluish-white Spica on 14 July.

Following behind is **Saturn**, in Libra. Shining at magnitude +0.6, the ringworld is setting about 1 am.

Neptune rises around 11 pm in Aquarius. At magnitude +7.8, you will need a telescope to spot the most distant planet. It's followed by **Uranus** (magnitude +5.8), rising about midnight in Pisces.

For early birds, brilliant **Venus** is visible from 3 am onwards, very low in the north-eastern dawn glow, at magnitude −3.8.

Just to the lower left of Venus – if you have a clear horizon – you may spot **Mercury**, from its greatest western elongation on 12 July through to about 25 July. During this period, the innermost planet brightens from magnitude +0.6 to −0.9.

Jupiter is too close to the Sun to be visible in July.

MOON

On 5 July, the First Quarter Moon is just to the right of Mars and Spica.

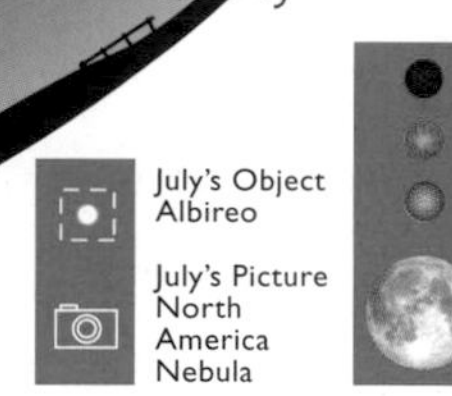

Mars
Saturn
Neptune
Moon

MOON		
Date	Time	Phase
5	12.59 pm	First Quarter
12	12.25 pm	Full Moon
19	3.08 am	Last Quarter
26	11.42 pm	New Moon

It lies near ringworld Saturn on 7 July. The red star below the Moon on 9 July is Antares. In the morning of 22 July, you'll find the crescent Moon above Aldebaran. In the pre-dawn sky of 24 July, a slender Moon lies to the right of Venus, with Mercury to the left of the Morning Star; there's even more of a challenge in the dawn twilight on 25 July, when the very thin Moon appears beneath Venus, and to the right of Mercury.

SPECIAL EVENTS

On **4 July**, at 1.13 am, the Earth reaches aphelion, its furthest point from the Sun – 152 million kilometres out.

The two brightest asteroids, **Ceres** (magnitude +8.5) and **Vesta** (magnitude +7.2), pass only 10 arcminutes apart on the night of **5/6 July**. It's the closest approach of these two minor planets – the twin targets of the DAWN space mission – since they were discovered over 200 years ago. With binoculars or a small telescope, you'll find these two tiny worlds in Virgo, 7 degrees above Mars.

JULY'S OBJECT

The constellation **Cygnus** represents a soaring swan, her wings outspread as she flies down the Milky Way. The lowest star in Cygnus, marking the swan's head, is **Albireo**. The name looks Arabic, but it actually has no meaning and is merely the result of errors in translation – from Greek, to Arabic, to Latin.

Binoculars reveal that Albireo is actually two stars in one. Use a telescope, and you'll be treated to one of the most glorious sky-sights – a dazzling yellow star teamed up with a blue companion.

Viewing tip
This is the month when you really need a good, unobstructed view to the southern horizon to make out the summer constellations of Scorpius and Sagittarius. They never rise high in temperate latitudes, so make the best of a southerly view – especially over the sea – if you're away on holiday. A good southern horizon is also best for views of the planets, because they rise highest when they're in the south.

The yellow star is a giant, near the end of its life. It's 80 times bigger than the Sun, and 1000 times brighter. The fainter blue companion is 'only' 200 times brighter than the Sun.

The spectacular colour contrast is due to the stars' different temperatures. The giant star is slightly cooler than our Sun, and shines with a yellowish glow. The smaller companion is far hotter: it's so incandescent that it shines not merely white-hot, but blue-white.

JULY'S PICTURE

Close to the star Deneb, the 'tail' of Cygnus (the Swan), the **North America Nebula** couldn't be more aptly named. Its shape looks uncannily like that continent! The glowing gas reveals a vast, active region of starbirth – while the dark 'Gulf of Mexico' (on the right of the image) is a cloud of stardust poised to create the next generations of the cosmos.

◀ Jim Coughlan captured this image of the North America Nebula – NGC 7000 – in North Devon. His telescope was a Takahashi 106 mm refractor, coupled with an Atik 383L+ mono CCD camera. Three narrowband filters were used – sulphur, hydrogen-alpha and oxygen – with a total exposure time of 1 hour per filter.

JULY'S TOPIC
Extrasolar planets

The tally of planets circling other stars now stands at nearly 1000 – with thousands more waiting to be confirmed.

The first came in 1995, when Swiss astronomers Michel Mayor and Didier Queloz discovered that the faint star 51 Pegasi – just to the right of the great Square of Pegasus – was being pulled backwards and forwards every four days. It had to be the work of a planet, tugging on its parent star. Astonishingly, this planet is around the same size as the Solar System's giant, Jupiter, but it is far closer to its star than Mercury. Astronomers call such planets 'hot jupiters'.

A team in California led by Geoff Marcy was already looking for planets, and soon found more. Now astronomers are finding whole solar systems with several worlds in stable orbits – like the six planets orbiting the star Gliese 667C. Three of these worlds are in the 'goldilocks zone' – possibly rocky planets not much more massive than the Earth, which may harbour liquid water.

The latest breakthroughs were made by NASA's Kepler orbiting spacecraft, which picked up tiny diminutions in light when a planet crossed the disc of its parent star. As we go to press, there are 2740 more candidates waiting to be checked out. Kepler researchers believe that the data they have garnered so far point to 'at least' 17 billion Earth-sized planets in our Galaxy.

This is the month when we usually say 'look out for shooting stars' – the famous **Perseid** meteors, which peak on the night of 12/13 August. This year, alas, visibility is dampened by the presence of the Moon. But there's a plus side: this month, we have a 'supermoon' – it appears bigger and brighter than our usual companion in space (see Special Events).

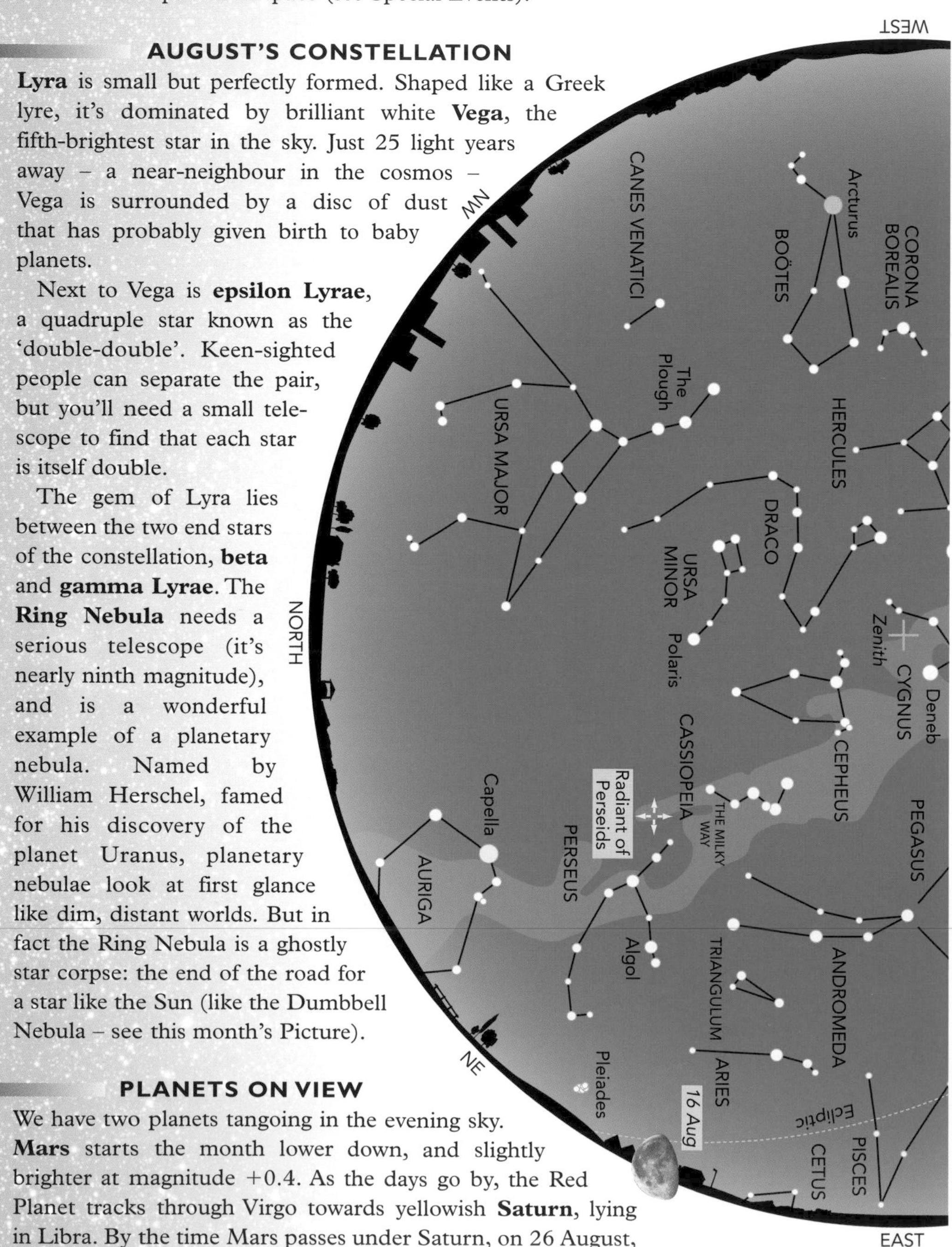

▼ *The sky at 11 pm in mid-August, with Moon positions at three-day intervals either side of Full Moon. The star positions are also correct for midnight*

AUGUST'S CONSTELLATION

Lyra is small but perfectly formed. Shaped like a Greek lyre, it's dominated by brilliant white **Vega**, the fifth-brightest star in the sky. Just 25 light years away – a near-neighbour in the cosmos – Vega is surrounded by a disc of dust that has probably given birth to baby planets.

Next to Vega is **epsilon Lyrae**, a quadruple star known as the 'double-double'. Keen-sighted people can separate the pair, but you'll need a small telescope to find that each star is itself double.

The gem of Lyra lies between the two end stars of the constellation, **beta** and **gamma Lyrae**. The **Ring Nebula** needs a serious telescope (it's nearly ninth magnitude), and is a wonderful example of a planetary nebula. Named by William Herschel, famed for his discovery of the planet Uranus, planetary nebulae look at first glance like dim, distant worlds. But in fact the Ring Nebula is a ghostly star corpse: the end of the road for a star like the Sun (like the Dumbbell Nebula – see this month's Picture).

PLANETS ON VIEW

We have two planets tangoing in the evening sky. **Mars** starts the month lower down, and slightly brighter at magnitude +0.4. As the days go by, the Red Planet tracks through Virgo towards yellowish **Saturn**, lying in Libra. By the time Mars passes under Saturn, on 26 August,

at the beginning of August, and 10 pm at the end of the month. The planets move slightly relative to the stars during the month.

it has dimmed almost to Saturn's magnitude of +0.7. The planetary pair is setting around 11 pm.

Neptune is at opposition on 29 August in Aquarius, and visible all night long. Though Neptune is closest to Earth this month, you need telescope power to track down this most distant planet, at a dim magnitude +7.8.

Its slightly nearer twin world, **Uranus** (magnitude +5.8), lies in Pisces and rises about 10 pm.

There's even closer planetary canoodling in the dawn sky. **Venus** is still the Morning Star, rising around 4 am and queening it over everything else, at magnitude −3.8. But **Jupiter** is now steaming upwards in the twilight glow; at magnitude −1.7, the giant planet is outshone only by Venus. The glorious pair makes a spectacular sight on the morning of 18 August, as they pass less than a quarter of a degree apart (see Special Events).

Mercury is lost in the Sun's glare throughout the month of August.

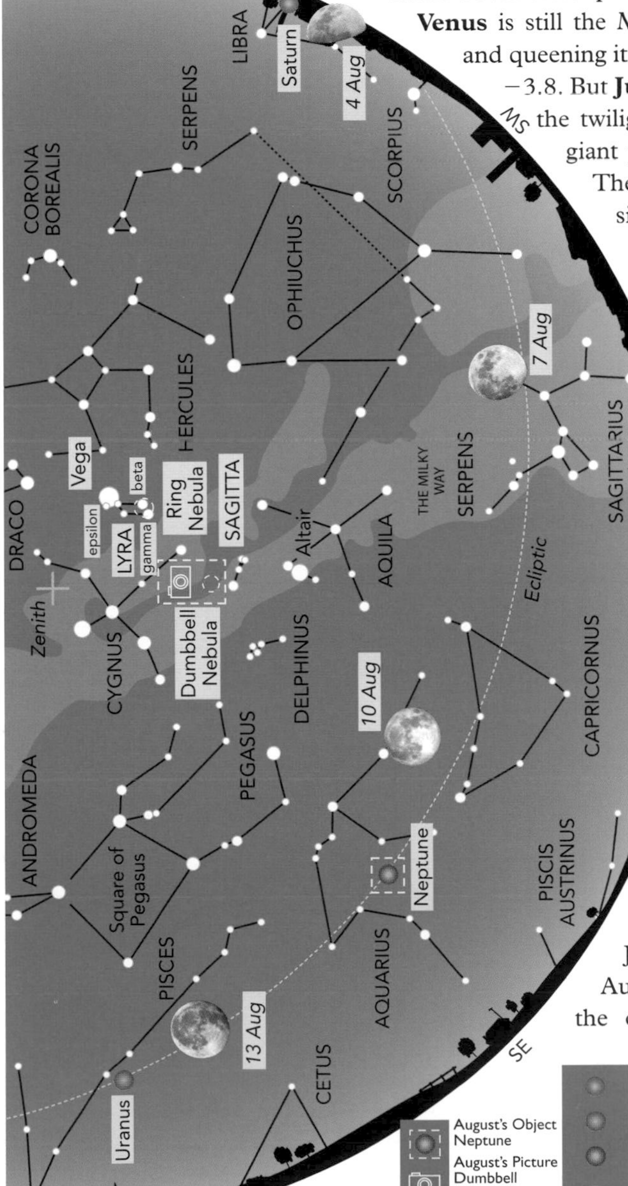

MOON

The crescent Moon lies between Mars and Spica on 2 August. On 3 August, it's between Saturn and Mars; while on 4 August, the First Quarter Moon is to the left of Saturn. You'll find the Moon near Antares on 5 August. On the night of 18/19 August, the Moon is to the left of Aldebaran. The waning crescent Moon lies just to the right of Venus and Jupiter on the morning of 23 August. Back in the evening sky, the crescent Moon is very close to

MOON		
Date	Time	Phase
4	1.49 am	First Quarter
10	7.09 pm	Full Moon
17	1.26 pm	Last Quarter
25	3.13 pm	New Moon

Saturn low in the south-west at dusk on 31 August, with Mars to the left.

SPECIAL EVENTS

We're due for a 'supermoon' on **10 August**, when the Moon is Full at its closest approach to the Earth this year, making it 14% bigger and 30% brighter than when our satellite is at its furthest. Contrary to popular rumour, the supermoon will not cause earthquakes, tsunamis or volcanic eruptions!

The maximum of the **Perseids** falls on **12/13 August**. It's normally among the best of the annual meteor showers; but, unfortunately, this year's strong moonlight will drown out all except the brightest shooting stars.

Just before dawn on **18 August**, the two brightest objects in the night sky (bar the Moon) – Venus and Jupiter – pass just 12 arcminutes from each other, as close as the Mizar–Alcor double star system in Ursa Major. With a small telescope, you can see Jupiter's globe and its moons along with the almost-full shape of Venus in the same field of view.

AUGUST'S OBJECT

With Pluto being demoted to a mere 'ice dwarf', **Neptune** is officially the most remote planet in our Solar System. It lies 4500 billion kilometres out – 30 times the Earth's distance from the Sun – in the twilight zone of our family of worlds, and takes nearly 165 years to complete one orbit.

Neptune is at its closest this year on 29 August, and just visible through a small telescope in Aquarius. But you need a space probe to get up close and personal to the gas giant planet. In 1989, Voyager 2 revealed a turquoise world 17 times heavier than Earth, cloaked in clouds of methane and ammonia.

◀ *With a Meade 127 mm refractor, Tom Howard captured this image of the Dumbbell Nebula from Crawley, West Sussex. He used a Nikon D7000 DSLR camera to photograph the nebula in 37 subexposures of 4 minutes each.*

👁 ***Viewing tip***
It's best to view your favourite objects when they're well clear of the horizon. When you look low down, you're seeing through a large thickness of the atmosphere – which is always shifting and turbulent. It's like trying to observe the outside world from the bottom of a swimming pool! This turbulence makes the stars appear to twinkle. Low-down planets also twinkle – although to a lesser extent, because they subtend tiny discs, and aren't so affected.

The most distant planet has a family of 13 moons, including Triton, which boasts erupting ice volcanoes. And it's encircled by very faint rings of dusty debris.

For a world so far from the Sun, Neptune is amazingly frisky. Its core blazes at nearly 5000°C – as hot as the Sun's surface. This internal heat drives dramatic storms, and winds of 2000 km/h the fastest in the Solar System.

AUGUST'S PICTURE

Located in the obscure constellation of Vulpecula (the Fox) – next to the head of diminutive **Sagitta** (the Arrow) – the **Dumbbell Nebula** is a huge hit with stargazers. At magnitude +7.5, it's visible through binoculars, and is a great target to capture on camera. This planetary nebula is the remains of an elderly star which puffed off its outer layers when its central nuclear reactor shut down – a fate that will befall our Sun (but not for 7 billion years!). At its centre is the biggest white dwarf star known – the dying, fuel-less core of the former red giant. The glowing shell of gas will waft away, and seed its environment to create a new generation of stars and planets.

AUGUST'S TOPIC
Centre of the Milky Way

The Milky Way stretches around our sky, as a gently glowing band. It's the inside view of a spiral-shaped galaxy of some 200–400 billion stars, with the Sun about halfway out. The centre of the Milky Way lies in the direction of Sagittarius, but the view – even for the most powerful optical telescopes – is blotted out by dense clouds of dark dust.

Now, telescopes observing infrared and radio waves have lifted the veil on the Galaxy's heart. They reveal stars and gas clouds whirling around at incredible speeds, up to 18 million km/h, in the grip of something with fantastically strong gravity. It's cast-iron evidence for a supermassive black hole, weighing as much as 4 million Suns.

When a speeding star or gas cloud comes too close to this invisible monster at the Galaxy's heart, it's ripped apart. There's a final shriek – a burst of radiation – before it falls into the black hole, and disappears from our Universe.

Around 10 million years ago, the black hole feasted on a huge cloud of gas, and 'burped' two giant bubbles of superhot gas that have been detected by satellites observing gamma rays. Currently, it's snacking on a smaller cloud that strayed too close last year. Invisible to our eyes it may be, but the galactic centre is where the action is!

This month, the nights become longer than the days, as the Sun migrates southwards in the sky (see Special Events). Autumn is here – with its unsettled weather – and we have wet star-patterns to match! **Aquarius** (the Water Carrier) is part of a group of aqueous star-patterns that includes **Cetus** (the Sea Monster), **Capricornus** (the Sea Goat), **Pisces** (the Fishes), **Piscis Austrinus** (the Southern Fish) and **Delphinus** (the Dolphin).

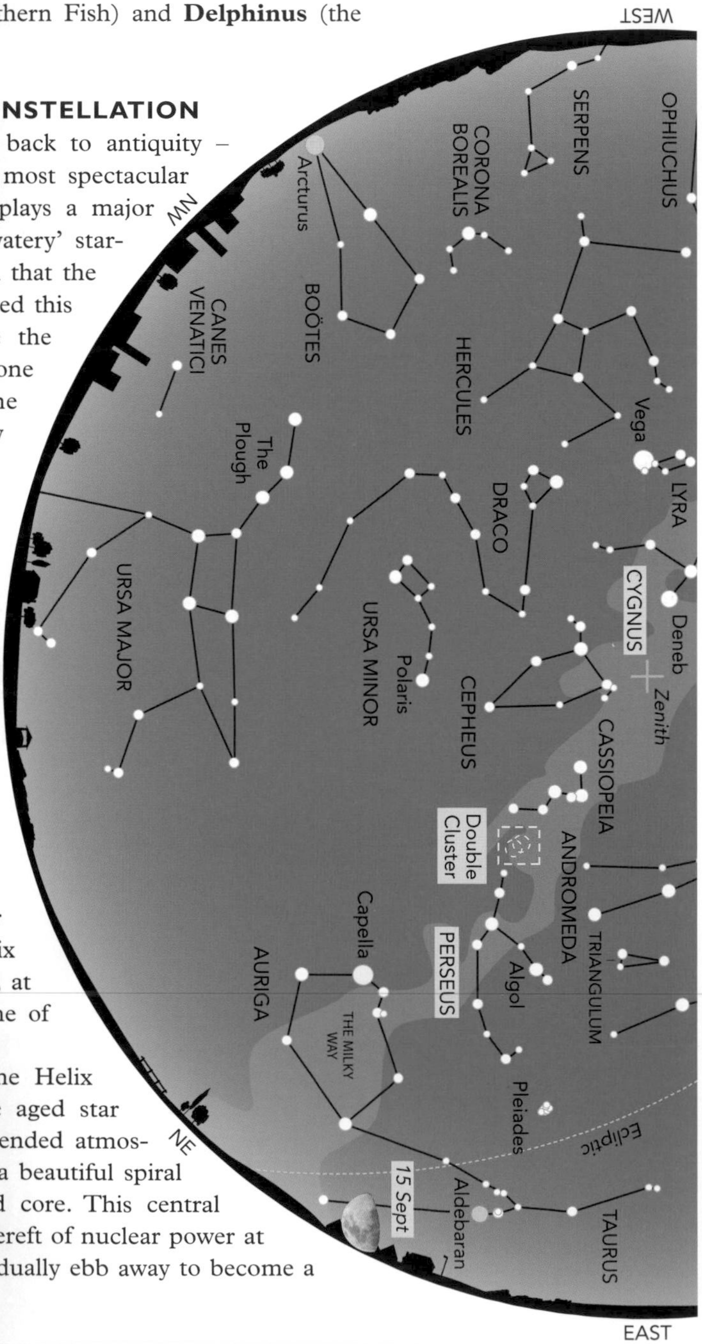

▼ The sky at 11 pm in mid-September, with Moon positions at three-day intervals either side of Full Moon. The star positions are also correct for midnight at

SEPTEMBER'S CONSTELLATION

With a pedigree stretching back to antiquity – although hardly one of the most spectacular constellations – **Aquarius** plays a major role among this season's 'watery' star-patterns. There's speculation that the ancient Babylonians associated this region with water because the Sun passed through this zone of the heavens during the rainy season, from February to March. They saw the faint, central four stars of Aquarius as a water jug, being poured by a man.

Aquarius boasts one of the most glorious sights in the sky in long-exposure images – the **Helix Nebula**. It's visible as a faint celestial ghost in binoculars or through a small telescope. Half the diameter of the Full Moon in the sky, the Helix Nebula is a star in its death throes. The Helix is a planetary nebula – and, at 700 light years away, it's one of the nearest known.

Once a red giant star, the Helix Nebula is the result of the aged star puffing off its unstable, distended atmosphere into space – forming a beautiful spiral shroud around its collapsed core. This central core is a white dwarf star: bereft of nuclear power at the end of its life, it will gradually ebb away to become a cold black cinder.

the beginning of September, and 10 pm at the end of the month. The planets move slightly relative to the stars during the month.

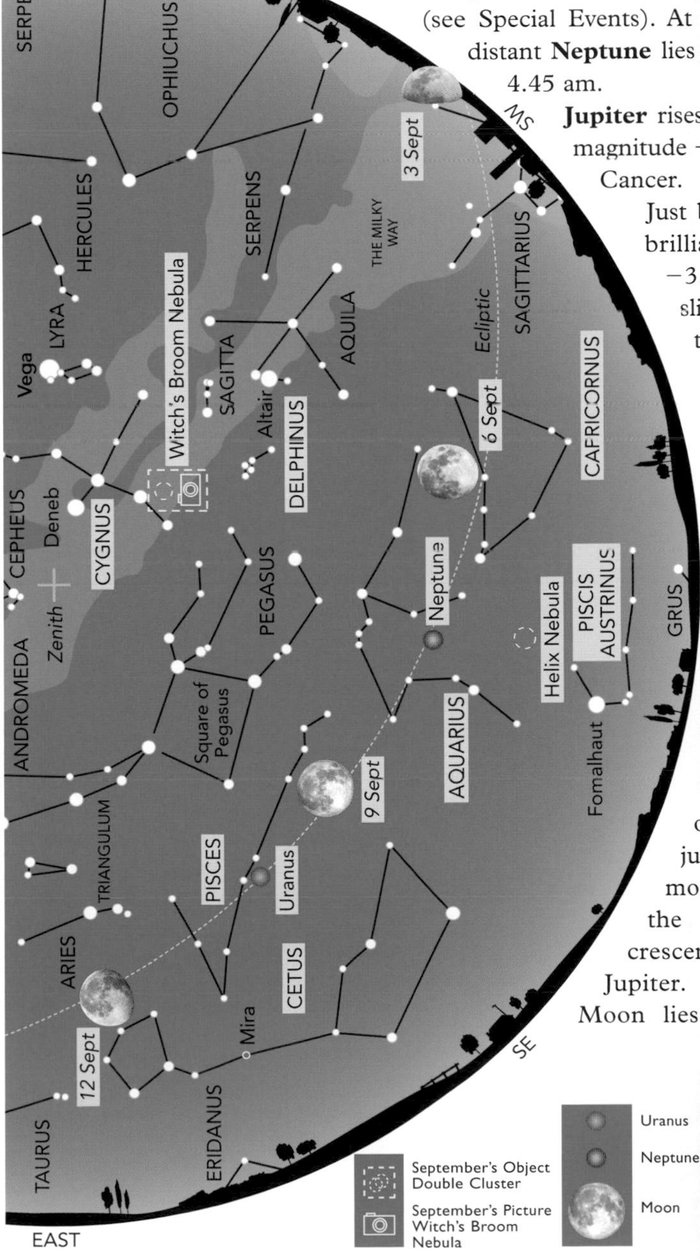

PLANETS ON VIEW

Mars and **Saturn** begin the month paired up in Libra, both at magnitude +0.7, and setting around 9 pm. But Mars is heading rapidly away from its more sedentary sibling, and by 13 September the Red Planet has entered Scorpius, passing slightly fainter Antares (magnitude +1.0) on 28 September (see Special Events).

You'll find **Uranus** (magnitude +5.7) in Pisces, rising just before 8 pm: the Moon is very close on 10/11 September (see Special Events). At magnitude +7.8, faint and distant **Neptune** lies in Aquarius, setting around 4.45 am.

Jupiter rises about 3 am, and shines at magnitude −1.7 in the dim constellation Cancer.

Just before dawn, you can catch brilliant **Venus**, at magnitude −3.8. The Morning Star is slipping downwards into the twilight glow, and now rises only an hour before the Sun.

Mercury is lost in the Sun's glare this month, even though it is at greatest eastern elongation on 21 September.

MOON

On the night of 10/11 September, the Moon grazes past Uranus (see Special Events). It lies in the Hyades star cluster on 14/15 September, passing just above Aldebaran. In the morning of 20 September, the brilliant 'star' near the crescent Moon is giant planet Jupiter. The waxing crescent Moon lies near Saturn on 27 and

MOON		
Date	Time	Phase
2	12.11 pm	First Quarter
9	2.38 am	Full Moon
16	3.05 am	Last Quarter
24	7.14 am	New Moon

28 September; it passes above Mars and Antares on 29 September.

> *Viewing tip*
> Now that the nights are drawing in earlier, and becoming darker, it's a good time to pick out faint, fuzzy objects such as star clusters, nebulae and galaxies. But don't even think about it near the time of Full Moon – its light will drown them out. The best time to observe 'deep-sky objects' is just before or after New Moon. Check the Moon phases timetables in the book.

SPECIAL EVENTS

The Moon passes extraordinarily close to Uranus on the night of **10/11 September**, with closest approach at 2.20 am: if you are observing from Shetland, you'll see the Moon actually hide the seventh planet. Take advantage of this rare alignment to locate the distant giant world, although the Moon is so bright you'll need binoculars or a telescope to spot Uranus.

It's the Autumn Equinox at 3.29 am on **23 September**. The Sun is over the Equator as it heads southwards in the sky, and day and night are equal.

On **28 September**, Mars passes Antares. It's an ideal chance to compare the colour of the Red Planet and the red giant star whose name means 'rival of Mars' – the hues are more obvious in binoculars.

SEPTEMBER'S OBJECT

Two objects this month: the beautiful **Double Cluster** in **Perseus**. These near-twin clusters of fledgling stars – each covering an area bigger than the Full Moon – are visible to the unaided eye and are a gorgeous sight in binoculars. They're loaded with glorious, young, blue supergiant stars, and are around 7500 light years away. Officially known as h and chi Persei, each cluster is a mere 12 million years old. Both contain several hundred stars, and are part of what's known as the Perseus OB1 Association – a loose group of bright, hot stars that were born at roughly the same time. Associations and star clusters are important to researchers. They allow astronomers to monitor stars that are the same age, but have different masses. The comparison helps researchers understand how stars evolve.

SEPTEMBER'S PICTURE

The **Witch's Broom Nebula** is the western rim of the vast Veil Nebula in **Cygnus**. This circular nebula, which – in its entirety – measures 50 light years across, was discovered by William Herschel (he who found Uranus) on 5 September 1784. He wrote of the Witch's Broom: 'Extended passes through 52 Cygni. Near two degrees in length.' The Veil Nebula is the remains of a star that exploded between 5000 and 8000 years ago. The expanding supernova remnant now covers a region 3 degrees across. It's relatively faint – but if you have a moderate telescope it's a great challenge to pick out the convoluted filaments of interstellar matter swept up by the shockwave from the exploding star.

▲ *This photograph was taken by Rob Hodgkinson from Weymouth, Dorset, with an 80 mm MB refractor in several sessions between July and September 2011. He used an Atik 16HR mono CCD camera through hydrogen-alpha, OIII and tricolour filters. The total exposure time was 11 hours and 22 minutes.*

SEPTEMBER'S TOPIC
Dark sky destinations

If you're frustrated about not being able to see the stars because of light pollution, take yourself off on an early autumn holiday to stargaze. It's warm enough to be outside, the nights are getting darker, and the bright stars of summer are still with us. Despite the rampant escalation of street-lighting, the UK still has some of the darkest spots in Europe where you can eyeball the heavens – as recognized by the International Dark Sky Association. The first to be picked out was Galloway Forest Dark Sky Park, in south-west Scotland. The next was a whole island – Sark Dark Sky Island, in the Channel Isles. The Association's third selection is Exmoor Dark Sky Reserve, which straddles Devon and Somerset. Plus, there's an extra bonus to visiting these havens of the heavens – all the locations are in fabulous countryside, where you can walk, explore, cycle and sightsee (not to mention visit some good pubs!). More details at www.darkskydiscovery.org.uk/dark_sky_places/.

The glories of October's skies can best be described as 'subtle'. The barren square of **Pegasus** dominates the southern sky, with **Andromeda** attached to his side. But there are two galaxies on view: the **Andromeda Galaxy** (see this month's Object) and its fainter cousin the **Triangulum Galaxy**. When you gaze upon these cosmic beauties, you are looking back over 2 million years in time.

▼ The sky at 11 pm in mid-October, with Moon positions at three-day intervals either side of Full Moon. The star positions are also correct for midnight at

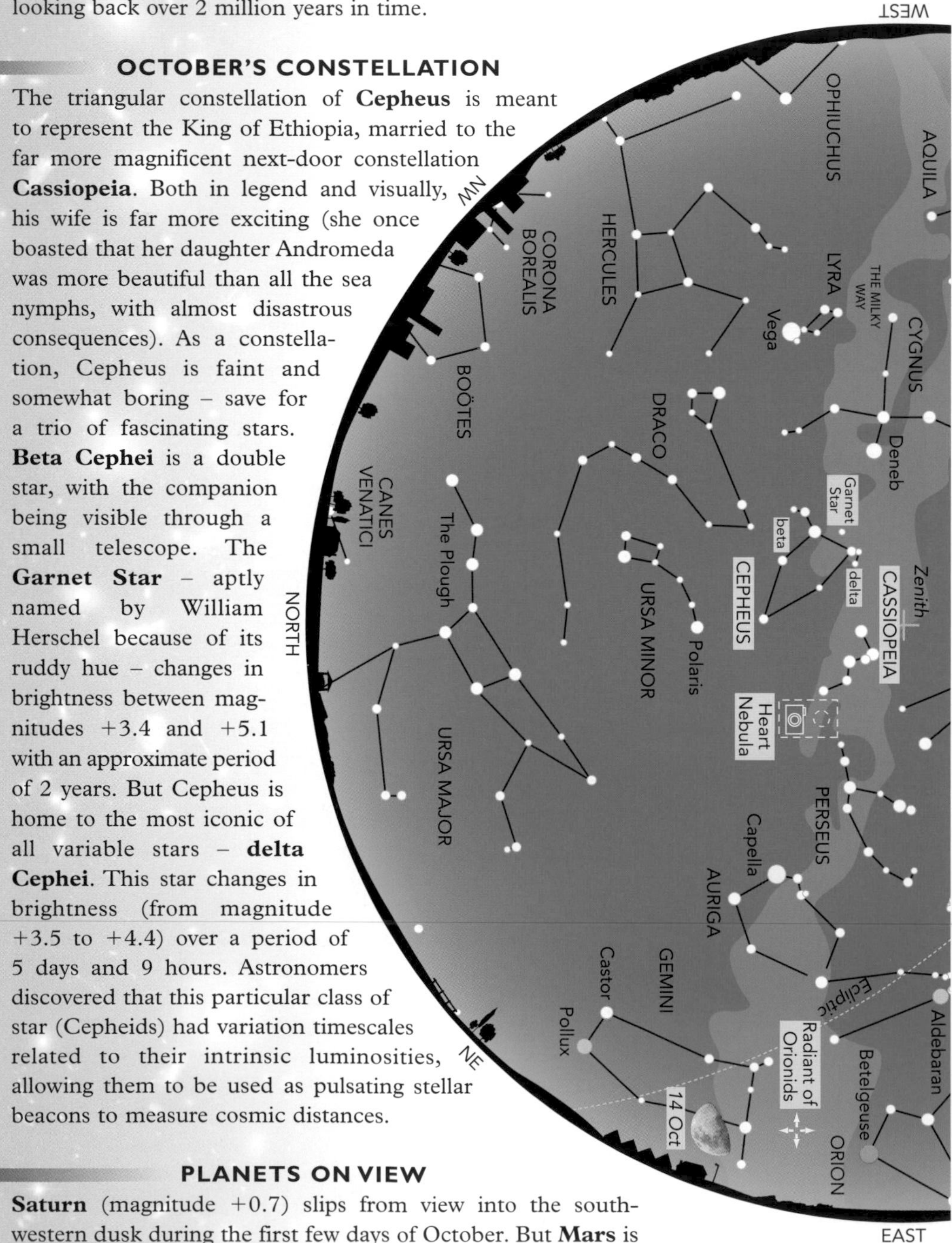

OCTOBER'S CONSTELLATION

The triangular constellation of **Cepheus** is meant to represent the King of Ethiopia, married to the far more magnificent next-door constellation **Cassiopeia**. Both in legend and visually, his wife is far more exciting (she once boasted that her daughter Andromeda was more beautiful than all the sea nymphs, with almost disastrous consequences). As a constellation, Cepheus is faint and somewhat boring – save for a trio of fascinating stars. **Beta Cephei** is a double star, with the companion being visible through a small telescope. The **Garnet Star** – aptly named by William Herschel because of its ruddy hue – changes in brightness between magnitudes +3.4 and +5.1 with an approximate period of 2 years. But Cepheus is home to the most iconic of all variable stars – **delta Cephei**. This star changes in brightness (from magnitude +3.5 to +4.4) over a period of 5 days and 9 hours. Astronomers discovered that this particular class of star (Cepheids) had variation timescales related to their intrinsic luminosities, allowing them to be used as pulsating stellar beacons to measure cosmic distances.

PLANETS ON VIEW

Saturn (magnitude +0.7) slips from view into the south-western dusk during the first few days of October. But **Mars** is

the beginning of October, and 9 pm at the end of the month (after the end of BST). The planets move slightly relative to the stars during the month.

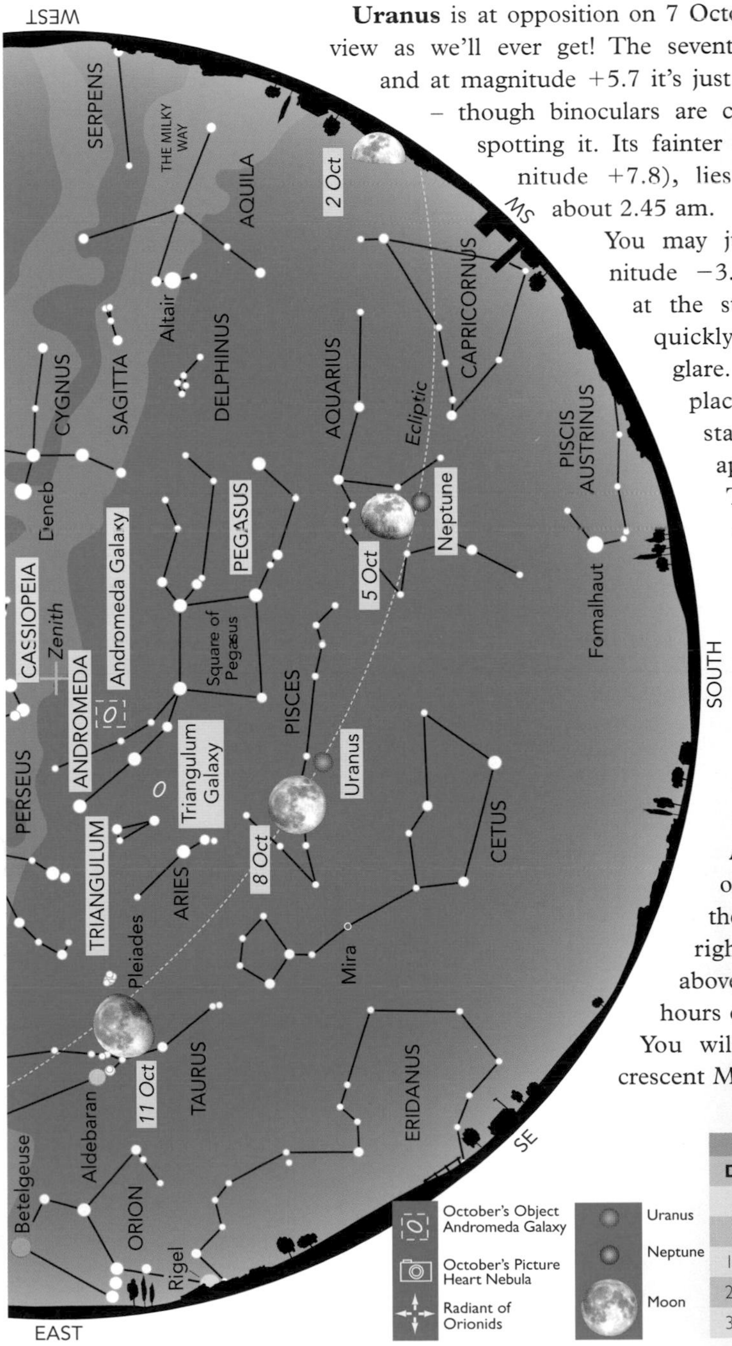

putting on a braver show: look for the Red Planet at magnitude +0.8, low in the south-west around 7 pm, as it moves through Ophiuchus – the '13th sign of the Zodiac' (see this month's Topic) – and into Sagittarius.

Giant planet **Jupiter** is rising around 1 am between Cancer and Leo, shining brighter than any star, at magnitude −1.8

Uranus is at opposition on 7 October, so this is as good a view as we'll ever get! The seventh planet lies in Pisces, and at magnitude +5.7 it's just visible to the naked eye – though binoculars are certainly a great help in spotting it. Its fainter sibling, **Neptune** (magnitude +7.8), lies in Aquarius and sets about 2.45 am.

You may just catch **Venus** (magnitude −3.8) in the morning sky at the start of October, but it quickly drops into the Sun's glare. Later in the month, its place is taken by **Mercury**, starting its best morning appearance of the year. The innermost planet emerges in the eastern dawn sky around 26 October, at magnitude +0.6, brightening to magnitude −0.3 by the end of the month.

MOON

On 11 and 12 October, the Moon is near Aldebaran. The mornings of 17 and 18 October see the crescent Moon to the right of Jupiter. The star above the Moon in the early hours of 19 October is Regulus. You will find Mars below the crescent Moon on 28 October.

MOON		
Date	Time	Phase
1	8.33 pm	First Quarter
8	11.50 am	Full Moon
15	8.12 pm	Last Quarter
23	10.57 pm	New Moon
31	2.48 am	First Quarter

SPECIAL EVENTS

The Pacific Ocean and the countries round its rim are treated to a total eclipse of the Moon on **8 October**, right next to Uranus: people either side of the Bering Strait will see the very rare sight of the planet being occulted by the eclipsed Moon. Unfortunately, nothing of this event is visible from the UK.

On **19 October**, a bright comet is due – not for us, but for Mars. Discovered at Siding Spring Observatory in New South Wales by Rob McNaught, Comet 2013A1 will zip past the Red Planet at a distance of a mere 120,000 kilometres. Expect stunning images from the robot spacecraft on Mars and in orbit.

Debris from Halley's Comet smashes into Earth's atmosphere on **21/22 October**, causing the annual **Orionid** meteor shower. This is a great year for viewing the shooting stars, as the Moon is well out of the way: best after midnight.

The Sun is partially eclipsed on **23 October**, as seen from most of North America and the north-east Pacific. As with the other three eclipses of 2014 (in this month and April), British viewers miss out entirely!

At 2 am on **26 October**, we see the end of British Summer Time for this year. Clocks go backwards by an hour.

OCTOBER'S OBJECT

Take advantage of autumn's newborn darkness to pick out one of our neighbouring galaxies – the **Andromeda Galaxy**, also known as M31 after its position in a catalogue of fuzzy patches compiled by Charles Messier. It covers an area four times bigger than the Full Moon. Like our Milky Way, it is a beautiful spiral shape, but – alas – it's presented to us almost edge-on.

The Andromeda Galaxy lies around 2.5 million light years away, and it's similar in size and shape to the Milky Way. It also hosts two bright companion galaxies – just like our Milky Way – as well as a flotilla of orbiting dwarf galaxies.

◄ *James McConnachie captured this image of the Heart Nebula from East Ayrshire, Scotland, with a modified Canon 400D DSLR through a William Optics 70 mm refractor. He used a combination of colour and hydrogen-alpha exposures, with a total exposure time of 4 hours 46 minutes.*

> 👁 ***Viewing tip***
> The Andromeda Galaxy is often described as the furthest object 'easily visible to the unaided eye'. But it's not *that* easy to see – especially if you are suffering from light pollution. The trick is to memorize the star patterns in Andromeda and look slightly to the *side* of where you expect the galaxy to be. This technique – called 'averted vision' – causes the image to fall on parts of the retina that are more light-sensitive than the central region, which is designed to see fine detail. It's worth bearing in mind when searching for any faint object in the sky.

Unlike other galaxies, which are receding from us (as a result of the expansion of the Universe), the Milky Way and Andromeda are approaching each other. It's estimated that they will merge in about 5 billion years' time. The result of the collision may be a colossal elliptical galaxy – nicknamed Milkomeda – which will be devoid of the gas that gives birth to stars, and dominated by ancient red giant stars.

OCTOBER'S PICTURE

The familiar W-shaped constellation of **Cassiopeia** is always on view in the sky in our northern latitudes, never being far from the pole. It contains some exotic denizens – such as the Cas A supernova remnant, the remains of a star that exploded some 300 years ago. But it has more subtle delights, such as the **Heart Nebula**, pictured here. Its shape says it all! Lying 7500 light years away, the nebula is a cradle to a cluster of newborn stars – Melotte 15 – which illuminate the gas cloud with their powerful radiation. Some of these stellar youngsters are 50 times heavier than the Sun.

OCTOBER'S TOPIC
The 13th sign of the Zodiac

For most of this month, Mars is chilling out in the constellation of Ophiuchus. But hang on – surely the Moon and planets usually sit around in the well-known signs of the Zodiac, like Leo, Gemini or Taurus? Not entirely true. When our familiar constellation patterns were drawn up in Mesopotamia and Greece over 2000 years ago, astronomers noticed that the Sun, Moon and planets kept to a distinct band in the sky. They divided this special band into the constellations of the Zodiac, and assigned one star-pattern for each month of the year. Misguided astrologers today still interpret the positions of the planets and the Sun in the Zodiac as omens for humankind – in particular, where the Sun is placed against the stars during your birth-month. But think you're a Gemini? You're likely to be a Taurus. The reason for this is that everything has slipped back since the Greeks. Due to the gravity of the Moon, the Earth's axis wobbles with a period of 26,000 years – a phenomenon called 'precession', which changes the positions of the Sun and planets relative to the background stars. In addition, when you look in detail at a sky-map, you'll find that this celestial highway crosses part of a 13th constellation – Ophiuchus. Alas: if you believe that you're a Sagittarian, you're probably really an Ophiuchian!

Now rising in the east, the beautiful star cluster of the **Pleiades** is a sure sign that winter is on the way. From Greece to Australia, ancient myths independently describe the stars as a group of young girls being chased by an aggressive male – often **Aldebaran** or **Orion**. Polynesian navigators used the Seven Sisters to mark the start of their year. 'A swarm of fireflies tangled in a silver braid' was the evocative description of the star cluster by Alfred, Lord Tennyson, in his 1842 poem 'Locksley Hall'.

▼ The sky at 10 pm in mid-November, with Moon positions at three-day intervals either side of Full Moon. The star positions are also correct for 11 pm at

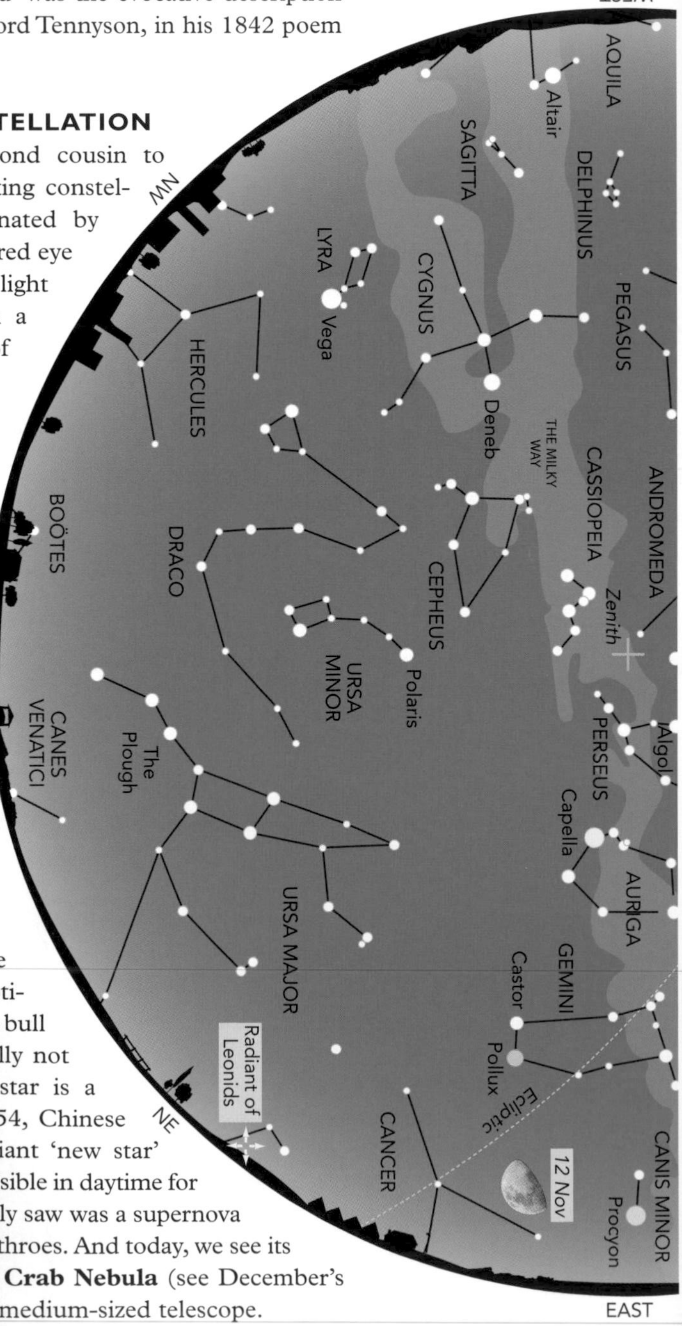

NOVEMBER'S CONSTELLATION

Taurus is very much a second cousin to brilliant **Orion**, but a fascinating constellation nonetheless. It's dominated by **Aldebaran**, the baleful blood-red eye of the celestial bull. Around 65 light years away, and shining with a (slightly variable) magnitude of +0.85, Aldebaran is a red giant star, but not one as extreme as neighbouring **Betelgeuse**. It is around the mass of the Sun. The 'head' of the bull is formed by the **Hyades** star cluster. The other famous star cluster in Taurus is the far more glamorous **Pleiades**, whose stars – although further away than the Hyades – are younger and brighter.

Taurus has two 'horns': the star **El Nath** (Arabic for 'the butting one') to the north, and **zeta Tauri** (whose Babylonian name Shurnarkabti-sha-shutu, meaning 'star in the bull towards the south', is thankfully not generally used!). Above this star is a stellar wreck – literally. In 1054, Chinese astronomers witnessed a brilliant 'new star' appear in this spot, which was visible in daytime for weeks. What the Chinese actually saw was a supernova – an exploding star in its death throes. And today, we see its still-expanding remains as the **Crab Nebula** (see December's Picture). It's visible through a medium-sized telescope.

the beginning of November, and 9 pm at the end of the month. The planets move slightly relative to the stars during the month.

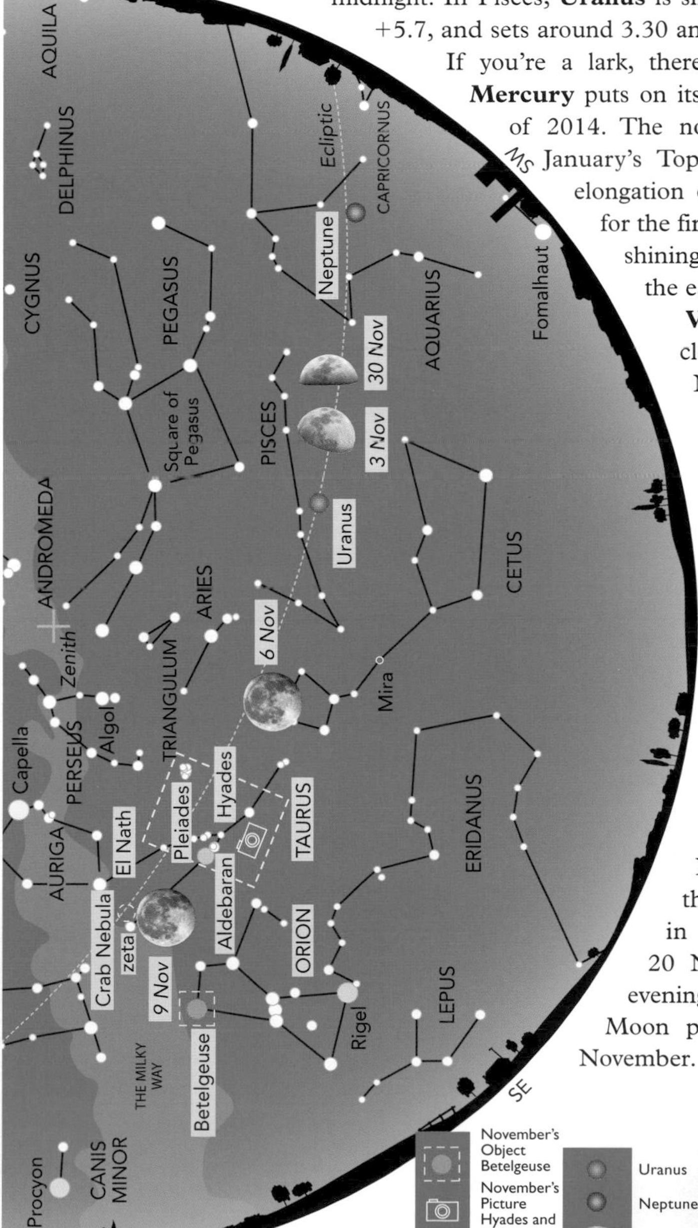

PLANETS ON VIEW

You'll find **Mars** low in the south-west, setting at 7.20 pm. The Red Planet shines at magnitude +1.0 in Sagittarius.

Jupiter is rising around 10.30 pm, on the border of Cancer and Leo. At magnitude −2.0, it's the brightest object in the night sky (apart from the Moon).

Neptune (magnitude +7.9) lies in Aquarius, setting about midnight. In Pisces, **Uranus** is slightly brighter at magnitude +5.7, and sets around 3.30 am.

If you're a lark, there's a real treat in store as **Mercury** puts on its best morning performance of 2014. The normally elusive planet (see January's Topic) is at greatest western elongation on 1 November. It's visible for the first three weeks of the month, shining at magnitude −0.6, low in the east around 6 am.

Venus and **Saturn** are too close to the Sun to be seen in November.

MOON

The Moon is very close to Uranus in the early evening of 4 November, though low in the twilight sky. The early evening of 8 November sees the Moon close to Aldebaran. On 13 November, the Moon lies near Jupiter; it forms a triangle with Jupiter and Regulus (to the left) on 14 November. The star near the waning crescent Moon in the morning of 19 and 20 November is Spica. In the evening sky, you'll find the crescent Moon passing above Mars on 26 November.

MOON		
Date	Time	Phase
6	10.23 pm	Full Moon
14	3.15 pm	Last Quarter
22	12.32 pm	New Moon
29	10.06 am	First Quarter

SPECIAL EVENTS

Around **10 November**, Philae – part of the Rosetta mission to Comet Churyumov–Gerasimenko – is due to become the first spacecraft to soft-land on a comet's nucleus. Philae will explore what this cosmic iceberg is made of, including its complement of organic compounds – the raw materials of life (see May's Topic).

The night of **17/18 November** sees the maximum of the **Leonid** meteor shower. You'll get the best views in the late evening, before the Moon rises around 1.30 am.

NOVEMBER'S OBJECT

Known to generations of schoolchildren as 'Beetle-Juice', **Betelgeuse** is one of the biggest stars known. If placed in the Solar System, it would swamp the planets all the way out to the asteroid belt. And it's one of just a few stars to be imaged as a visible disc from Earth.

Almost 1000 times wider than the Sun, Betelgeuse is a serious red giant – a star close to the end of its life. Its middle-aged spread has been created by the intensifying nuclear reactions

◄ *From Flackwell Heath in Buckinghamshire, Robin Scagell photographed the two clusters in Taurus with a diffusion filter to emphasize the colour differences between the stars. He used a combination of two exposures totalling 24 minutes, with a 55 mm lens on a Canon 10D at ISO 200 through a diffusing filter to bring out the star colours – which digital imaging often fails to reproduce.*

> **Viewing tip**
> If you want to stargaze at this most glorious time of year, dress up warmly! Lots of layers are better than a heavy coat as they trap air next to your skin – and heavy-soled boots stop the frost creeping up your legs. It may sound anorakish, but a woolly hat really does stop one-third of your body heat escaping through the top of your head. And – alas – no hipflask of whisky. Alcohol constricts the veins and makes you feel even colder.

in its core, causing its outer layers to swell and cool. The star also fluctuates slightly in brightness as it tries to get a grip on its billowing gases.

The star's prominence has led to it attracting several names – including 'The Armpit of the Sacred One'!

Betelgeuse will exit the Universe in a spectacular supernova explosion. As a result of the breakdown of nuclear reactions at its heart, the star will explode – and it will shine as brightly in our skies as the Moon.

NOVEMBER'S PICTURE

With **Taurus** making its presence felt this month, we've homed in on its two star clusters – the **Hyades** (at lower left) and the **Pleiades** (at upper right). This fantastic image brings out the differences in the star colours between the clusters – and the distribution of their stars. The V-shaped Hyades (our closest star cluster) is over 600 million years old, and its stars are starting to lose the dazzling blue of youth. They are also beginning to make their own individual paths across the sky. (Incidentally, brilliant red giant Aldebaran – seen at left in the image – is an interloper, and actually lies much closer than the Hyades). Contrast the Hyades – a rather dispersed, dull-looking cluster – with the Pleiades. These brilliant, hot young blue stars are a mere 75–100 million years old, and are still tightly bound by the gravity that created them.

NOVEMBER'S TOPIC
Cosmic blasts

At 9.20 in the morning of 15 February 2013, citizens of the Russian city of Chelyabinsk thought that the end of the world had come. Out of nowhere, there appeared a brilliant fireball, which exploded 20 kilometres above the heads of the horrified residents. The shockwave caused by the blast – which was 20 times more powerful than the atomic bombs aimed at Hiroshima and Nagasaki – blew in windows, injuring some 1500 people with flying glass. Was it a nuclear missile? No: it was a 10,000-tonne meteorite, which had strayed from the asteroid belt. The event was a chilling reminder that we have to take the asteroid threat seriously. The Earth continues to be bombarded with debris left over from the formation of the Solar System. And, every so often, a lump of cosmic rock far bigger than the Chelyabinsk meteorite hits our planet – as the dinosaurs learned to their cost 65 million years ago. Using ever bigger telescopes and new satellites, astronomers are now redoubling their efforts to detect – and perhaps even control – rogue asteroids.

The glorious constellations of winter are riding high in the sky. **Orion**, with his hunting dogs **Canis Major** and **Canis Minor**, is dominating the heavens, fighting his adversary **Taurus** (the Bull). This may be the darkest month, but we're treated to a welcome fireworks display on 13 December as the **Geminid** meteors streak into our atmosphere.

▼ The sky at 10 pm in mid-December, with Moon positions at three-day intervals either side of Full Moon. The star positions are also correct for 11 pm at

DECEMBER'S CONSTELLATION

You can't ignore **Gemini** in December. High in the south-east, the constellation is crowned by the stars **Castor** and **Pollux**, which are of similar brightness and represent the heads of a pair of twins – their stellar bodies run in parallel lines of stars towards the west. Legend has it that Castor and Pollux were twins, conceived on the same night by the princess Leda. On the night she married the King of Sparta, wicked old Zeus (Jupiter) also invaded the marital suite, disguised as a swan. Pollux was the result of the liaison with Jupiter – and therefore immortal – while Castor was merely a human being. But the pair were so devoted to each other that Zeus decided to grant Castor honorary immortality, and placed both him and Pollux amongst the stars.

Castor is an amazing star: it's part of a family of six. Even through a small telescope, you can see that Castor is a double star, comprising two stars circling each other. Both of these are double (although you need special equipment to detect this). Then there's also another outlying star, visible through a telescope, which also turns out to be double.

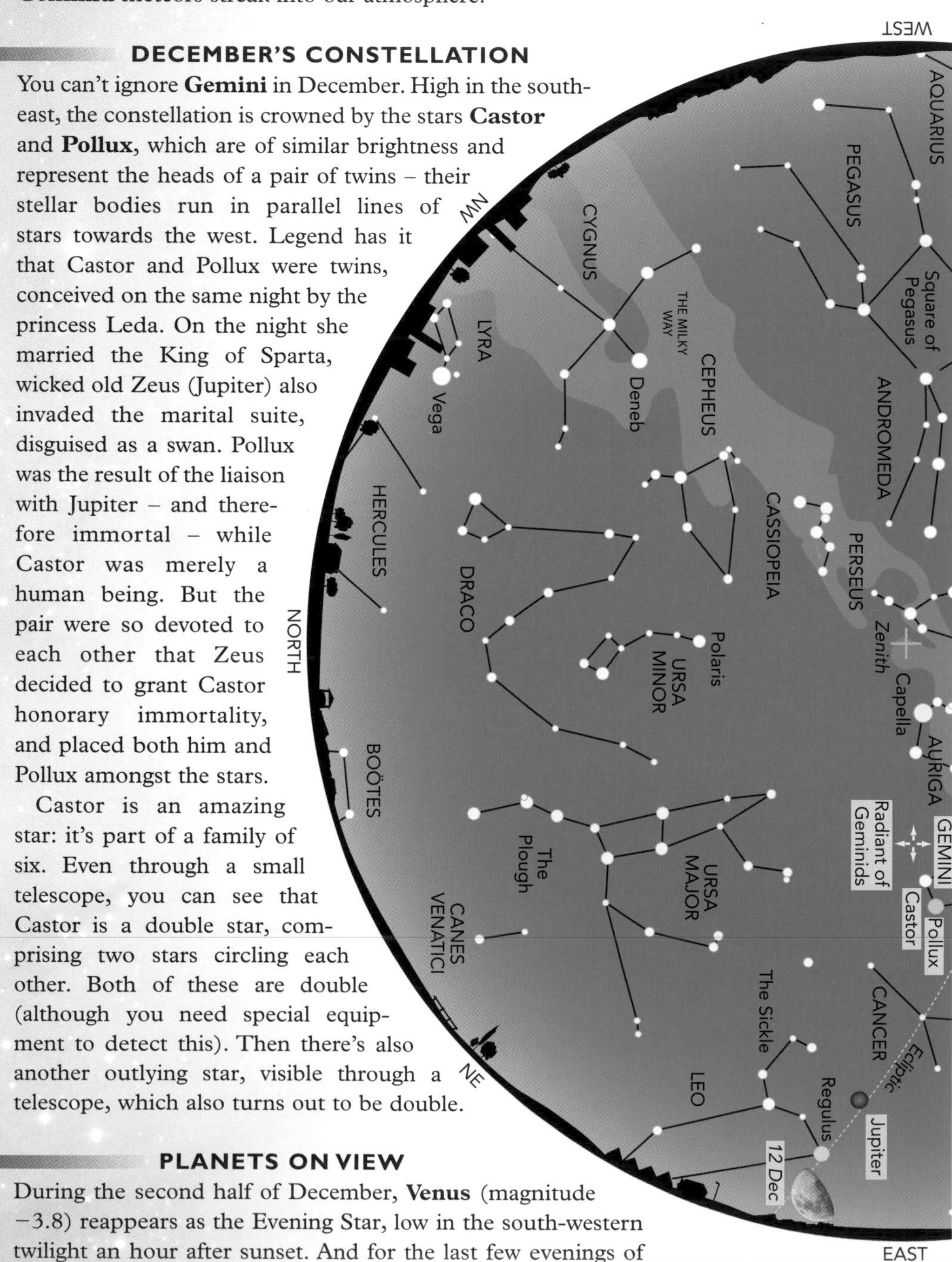

PLANETS ON VIEW

During the second half of December, **Venus** (magnitude −3.8) reappears as the Evening Star, low in the south-western twilight an hour after sunset. And for the last few evenings of

the beginning of December, and 9 pm at the end of the month. The planets move slightly relative to the stars during the month.

the year, you may spot **Mercury** just to its lower right; it's some 20 times fainter than Venus, at magnitude −0.7, but should be an easy catch in binoculars.

Mars is hanging on in the evening sky, as it has been all year, and is now setting at 7.30 pm. At magnitude +1.0, the Red Planet lies in Capricornus.

Around 9 pm, brilliant **Jupiter** is rising in the east, between Cancer and Leo. The giant planet blazes at magnitude −2.2.

Over in Aquarius, **Neptune** (magnitude +7.9) is setting about 10 pm. And **Uranus**, in Pisces at magnitude +5.8, slips below the horizon at around 1.30 am.

Saturn is now appearing in the morning sky: by the end of December, you can spot it rising at 5 am. At magnitude +0.7, the ringed planet lies in Libra.

MOON

The Moon once again passes very close to Uranus, on the night of 1/2 December. On the night of 5/6 December, the Full Moon passes through the Hyades and skims past Aldebaran. It is near Jupiter and Regulus on 11 and 12 December. Spica is the star near the crescent Moon in the morning of 17 December. Before dawn on 19 and 20 December, you'll find the slender Moon near Saturn. In the evening sky, the narrowest crescent Moon reappears on 23 December, above Venus. The Moon is near Mars on 24 and 25 December.

MOON

Date	Time	Phase
6	12.27 pm	Full Moon
14	12.51 pm	Last Quarter
22	1.36 am	New Moon
28	6.31 pm	First Quarter

SPECIAL EVENTS

The maximum of the **Geminid** meteor shower falls on **13/14 December**. These shooting stars are debris shed from an asteroid called Phaethon and are therefore quite substantial – and hence bright.

The Winter Solstice occurs at 11.03 pm on **21 December**. As a result of the tilt of Earth's axis, the Sun reaches its lowest point in the heavens as seen from the northern hemisphere: we get the shortest days, and the longest nights.

DECEMBER'S OBJECT

The **Pleiades** star cluster is one of the most familiar sky-sights. Though it's well known as the Seven Sisters, skywatchers see any number of stars but seven! Most people can pick out the six brightest stars, while very keen-sighted observers can discern up to 11 stars. These are just the most luminous in a group of at least 1000 stars, lying about 400 light years away (although there's an ongoing debate about the precise distance!). The brightest stars in the Pleiades are hot and blue (see November's

▼ Alan Clitherow took this picture of the Crab Nebula from Fife, Scotland, using a SkyWatcher MN190P Maksutov-Newtonian of f/5.1 and 1000 mm focal length. The camera was an Atik 314L monochrome CCD, using two narrowband filters, a Baader 35 nm-pass hydrogen-alpha filter and an Astronomik 11 nm-pass oxygen-3 filter, mapped to red and blue respectively. Ten exposures of 10 minutes each per filter.

Picture), and all the stars are young – estimated to be 75–100 million years old. They were born together, and have yet to go their separate ways. The fledgling stars have blundered into a cloud of gas in space, which looks like gossamer on webcam images. Even to the unaided eye or through binoculars, they are still a beautiful sight.

Viewing tip
This is the month when you may be thinking of buying a telescope as a Christmas present for a budding stargazer. Beware! Unscrupulous websites and mail-order catalogues selling 'gadgets' often advertise small telescopes that boast huge magnifications. This is known as 'empty magnification' – blowing up an image that the lens or mirror simply doesn't have the ability to get to grips with, so all you see is a bigger blur. A rule of thumb is to use a maximum magnification no greater than twice the diameter of your lens or mirror in millimetres. So if you have a 100 mm reflecting telescope, go no higher than 200× magnification.

DECEMBER'S PICTURE

Another Taurus image! This time, it's the iconic **Crab Nebula**: the remains of a star that was seen to explode by the Chinese in AD 1054, and was visible for almost two years afterwards. The supernova remnant – around 10 light years across – is a wreck of twisted filaments. At its heart is the collapsed core of the dead star – a fiercely-spinning neutron star, known as a pulsar, which shoots out beams of light and radio waves into space. The pulsar's energy lights up the nebula, making it shine a ghostly blue.

DECEMBER'S TOPIC
Midwinter megalithic monuments

Today's Druids have made it easy for themselves – they celebrate at Stonehenge in the warmth of June. But maybe the builders of this ancient edifice were actually celebrating chilly Midwinter's Day! Instead of standing inside the stone circle, watching the summer Sun rising over the outlying Heel Stone, the Sun Chief was perhaps stationed at the Heel Stone, observing the pale winter Sun setting through the stone arches of the massive monument.

No one can know what these prehistoric farmers were thinking, of course. But when we visited another of the world's ancient observatories, at Chaco Canyon in New Mexico, USA, we discovered not just mute stones, but a living tradition in the surrounding villages where the Sun Chiefs' main rites take place at the Winter Solstice.

It makes sense. Early people were worried by the Sun progressively departing from the sky during autumn. Would it disappear altogether? Midwinter was a critical time, when the Sun turned round, days became longer and eventually warmth reappeared.

In the British Isles, we find ancient burial mounds at Newgrange in Ireland and Maes Howe in Orkney, where the Sun shines straight up a subterranean passage to the heart of the mound on Midwinter's Day.

Our money, then, is on Stonehenge being a midwinter monument, too. Forget the Druids: be there for sunset on 21 December!

There's always something to see in our Solar System, from planets to meteors or the Moon. These objects are very close to us – in astronomical terms – so their positions, shapes and sizes appear to change constantly. It is important to know when, where and how to look if you are to enjoy exploring Earth's neighbourhood. Here we give the best dates in 2014 for observing the planets and meteors (weather permitting!), and explain some of the concepts that will help you to get the most out of your observing.

Magnitudes

Astronomers measure the brightness of stars, planets and other celestial objects using a scale of *magnitudes*. Somewhat confusingly, fainter objects have higher magnitudes, while brighter objects have lower magnitudes; the most brilliant stars have negative magnitudes! Naked-eye stars range from magnitude −1.5 for the brightest star, Sirius, to +6.5 for the faintest stars you can see on a really dark night.

As a guide, here are the magnitudes of selected objects:

Sun	−26.7
Full Moon	−12.5
Venus (at its brightest)	−4.7
Sirius	−1.5
Betelgeuse	+0.4
Polaris (Pole Star)	+2.0
Faintest star visible to the naked eye	+6.5
Faintest star visible to the Hubble Space Telescope	+31

THE INFERIOR PLANETS

A planet with an orbit that lies closer to the Sun than the orbit of Earth is known as *inferior*. Mercury and Venus are the inferior planets. They show a full range of phases (like the Moon) from the thinnest crescents to full, depending on their position in relation to the Earth and the Sun. The diagram below shows the various positions of the inferior planets. They are invisible when at *conjunction*, when they are either behind the Sun, or between the Earth and the Sun, and lost in the latter's glare.

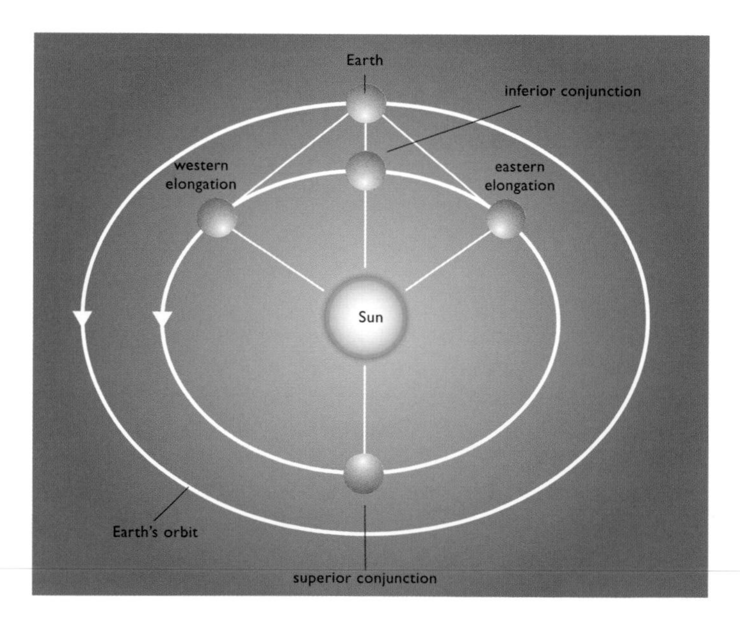

◀ *At eastern or western elongation, an inferior planet is at its maximum angular distance from the Sun. Conjunction occurs at two stages in the planet's orbit. Under certain circumstances, an inferior planet can transit across the Sun's disc at inferior conjunction.*

Mercury

Mercury has two great evening appearances: January–February, and again in May. The September evening apparition is completely lost in the bright twilight sky, but the planet is visible again during the last few days of the year. In contrast, the morning appearances get better during 2014. We're not going to see Mercury for its March apparition, but it's visible in the dawn skies of July; and we can look forward to the planet's best morning appearance in October–November.

Maximum elongations of Mercury in 2014

Date	Separation
31 January	18° east
14 March	28° west
25 May	23° east
12 July	21° west
21 September	26° east
1 November	19° west

Maximum elongation of Venus in 2014	
Date	Separation
22 March	47° west

Venus

Venus bookends 2014 as the Evening Star, during the first few days of January and again in December. Otherwise, it's in the morning sky all year, most prominent in February and March, when it rises from the dawn glow into a darker sky.

THE SUPERIOR PLANETS

The superior planets are those with orbits that lie beyond that of the Earth. They are Mars, Jupiter, Saturn, Uranus and Neptune. The best time to observe a superior planet is when the Earth lies between it and the Sun. At this point in a planet's orbit, it is said to be at *opposition*.

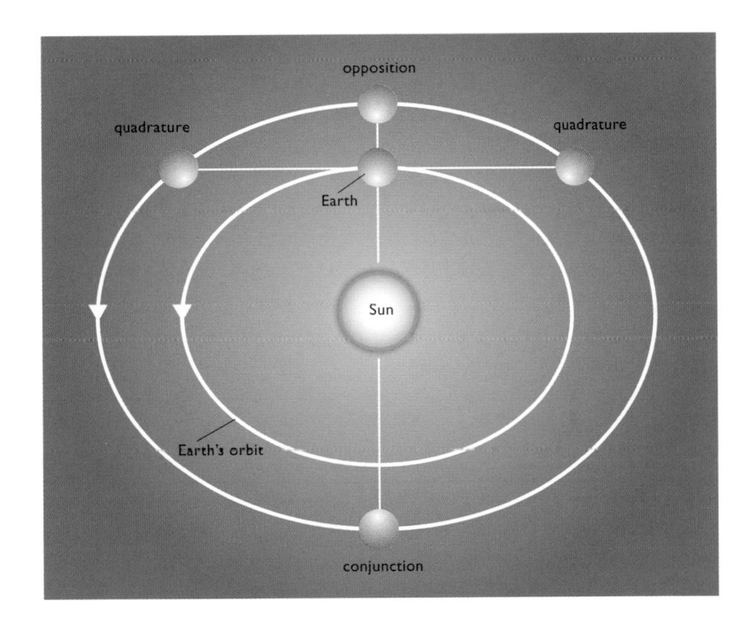

► *Superior planets are invisible at conjunction. At quadrature the planet is at right angles to the Sun as viewed from Earth. Opposition is the best time to observe a superior planet.*

Progress of Mars through the constellations	
January–July	Virgo
August–mid September	Libra
Late September	Scorpius
October	Ophiuchus
November	Sagittarius
December	Capricornus

Mars

The Red Planet is visible all year, reaching opposition on 8 April. At the start of 2014, it rises at midnight, but by April it's visible all night long. From late summer until the end of the year, you'll find the planet hanging around low in the evening sky.

Jupiter

The giant planet is at opposition on 5 January, when it's visible all night long. Jupiter is brilliant in the evening sky through to June, in Gemini; then it reappears in the morning sky in August, passing through Cancer towards Leo. By November, you'll catch Jupiter in the evening sky again.

Saturn

You'll find the ringed planet in Libra all year. Saturn is prominent in the morning sky at the start of 2014, and is visible all night when it reaches opposition on 10 May, remaining a feature of the evening sky until October. Saturn reappears in the morning sky in December.

Uranus

Just perceptible to the naked eye, Uranus is visible from January to March, and then from July to December. It resides in Pisces all year, and is at opposition on 7 October.

Neptune

Lying in Aquarius all year, the most distant planet is at opposition on 29 August. Neptune can be seen – though only through a telescope – in January, and then from July to the end of the year.

Astronomical distances

For objects in the Solar System, such as the planets, we can give their distances from the Earth in kilometres. But the distances are just too huge once we reach out to the stars. Even the nearest star (Proxima Centauri) lies 25 million million kilometres away.

So astronomers use a larger unit – the *light year*. This is the distance that light travels in one year, and it equals 9.46 million million kilometres.

Here are the distances to some familiar astronomical objects, in light years:

Proxima Centauri	4.2
Betelgeuse	640
Centre of the Milky Way	27,000
Andromeda Galaxy	2.5 million
Most distant galaxies seen by the Hubble Space Telescope	13 billion

SOLAR AND LUNAR ECLIPSES

Solar Eclipses

There are two solar eclipses in 2014, neither visible from the UK. On 29 April, parts of Antarctica are treated to an annular solar eclipse, which is visible as a partial eclipse across Australia and the southern Indian and Atlantic Oceans. The partial solar eclipse of 23 October can be seen from much of North America and the north-east Pacific: from northern Canada, 80% of the Sun is covered by the Moon.

Lunar Eclipses

The year 2014 has two lunar eclipses, neither visible from the UK. There's a total eclipse on 15 April, which can be seen from the whole of the Americas and the Pacific Ocean, and also the western Atlantic Ocean. The total lunar eclipse of 8 October is visible from most of North America (which will be treated to a partial solar eclipse two weeks later), plus the Pacific, Australia, New Zealand and eastern Asia.

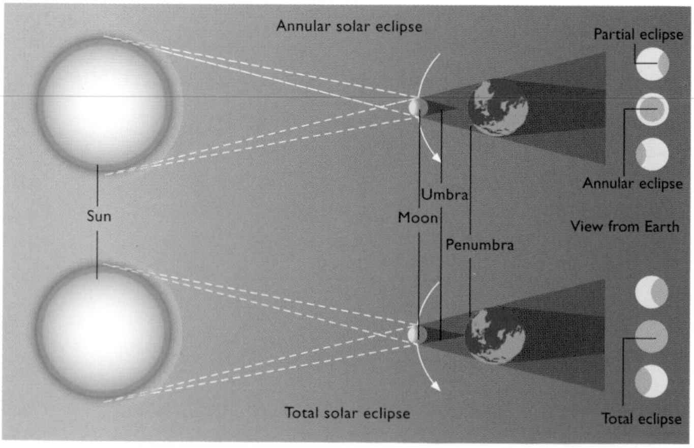

◀ *Where the dark central part (the umbra) of the Moon's shadow reaches the Earth, we see a total eclipse. People located within the penumbra see a partial eclipse. If the umbral shadow does not reach the Earth, we see an annular eclipse. This type of eclipse occurs when the Moon is at a distant point in its orbit and is not quite large enough to cover the whole of the Sun's disc.*

Dates of maximum for selected meteor showers	
Meteor shower	Date of maximum
Quadrantids	3/4 January
Lyrids	21/22 April
Eta Aquarids	5/6 May
Perseids	12/13 August
Orionids	21/22 October
Leonids	17/18 November
Geminids	13/14 December

METEOR SHOWERS

Shooting stars – or *meteors* – are tiny particles of interplanetary dust, known as *meteoroids*, burning up in the Earth's atmosphere. At certain times of year, the Earth passes through a stream of these meteoroids (usually debris left behind by a comet) and we see a *meteor shower*. The point in the sky from which the meteors appear to emanate is known as the *radiant*. Most showers are known by the constellation in which the radiant is situated. In 2014, as well as the regular showers, we may be treated to an outburst of meteors from newly discovered Comet LINEAR on 23/24 May (see May's Special Events).

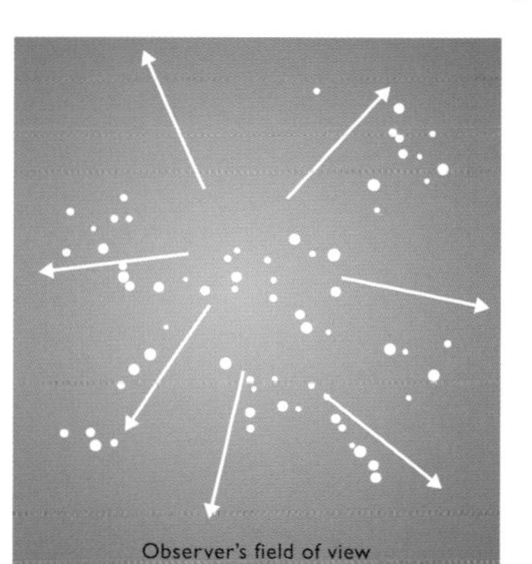

► *Meteors from a common source, occurring during a shower, enter the atmosphere along parallel trajectories. As a result of perspective, however, they appear to diverge from a single point in the sky – the radiant.*

When watching meteors for a co-ordinated meteor programme, observers generally note the time, seeing conditions, cloud cover, their own location, the time and brightness of each meteor, and whether it was from the main meteor stream. It is also worth noting details of persistent afterglows (trains) and fireballs, and making counts of how many meteors appear in a given period.

Angular separations

Astronomers measure the distance between objects, as we see them in the sky, by the angle between the objects in degrees (symbol °). From the horizon to the point above your head is 90 degrees. All around the horizon is 360 degrees.

You can use your hand, held at arm's length, as a rough guide to angular distances, as follows:

Width of index finger 1°
Width of clenched hand 10°
Thumb to little finger on outspread hand 20°

For smaller distances, astronomers divide the degree into 60 arcminutes (symbol ′), and the arcminute into 60 arcseconds (symbol ″).

COMETS

Comets are small bodies in orbit about the Sun. Consisting of frozen gases and dust, they are often known as 'dirty snowballs'. When their orbits bring them close to the Sun, the ices evaporate and dramatic tails of gas and dust can sometimes be seen.

A number of comets move round the Sun in fairly small, elliptical orbits in periods of a few years; others have much longer periods. Most really brilliant comets have orbital periods of several thousands or even millions of years. The exception is Comet Halley, a bright comet with a period of about 76 years. It was last seen with the naked eye in 1986.

Binoculars and wide-field telescopes provide the best views of comet tails. Larger telescopes with a high magnification are necessary to observe fine detail in the gaseous head (*coma*). Most comets are discovered with professional instruments, but a few are still found by experienced amateur astronomers.

If Comet ISON lives up to prediction, you should be able to see it with the naked eye at the start of January, but by February it fades to telescopic visibility. None of other known comets will reach naked-eye brightness this year, but another bright new comet may put in a surprise appearance.

Deep-sky objects are 'fuzzy patches' that lie outside the Solar System. They include star clusters, nebulae and galaxies. To observe the majority of deep-sky objects you will need binoculars or a telescope, but there are also some beautiful naked-eye objects, notably the Pleiades and the Orion Nebula.

The faintest object that an instrument can see is its *limiting magnitude*. The table gives a rough guide, for good seeing conditions, for a variety of small- to medium-sized telescopes.

We have provided a selection of recommended deep-sky targets, together with their magnitudes. Some are described in more detail in our monthly 'Object' features. Look on the appropriate month's map to find which constellations are on view, and then choose your objects using the list below. We have provided celestial coordinates for readers with detailed star maps. The suggested times of year for viewing are when the constellation is highest in the sky in the late evening.

Limiting magnitude for small to medium telescopes

Aperture (mm)	Limiting magnitude
50	+11.2
60	+11.6
70	+11.9
80	+12.2
100	+12.7
125	+13.2
150	+13.6

RECOMMENDED DEEP-SKY OBJECTS

Andromeda – autumn and early winter	
M31 (NGC 224) Andromeda Galaxy	3rd-magnitude spiral galaxy *RA 00h 42.7m Dec +41° 16′*
M32 (NGC 221)	8th-magnitude elliptical galaxy, a companion to M31 *RA 00h 42.7m Dec +40° 52′*
M110 (NGC 205)	8th-magnitude elliptical galaxy *RA 00h 40.4m Dec +41° 41′*
NGC 7662 Blue Snowball	8th-magnitude planetary nebula *RA 23h 25.9m Dec +42° 33′*
Aquarius – late autumn and early winter	
M2 (NGC 7089)	6th-magnitude globular cluster *RA 21h 33.5m Dec –00° 49′*
M72 (NGC 6981)	9th-magnitude globular cluster *RA 20h 53.5m Dec –12° 32′*
NGC 7293 Helix Nebula	7th-magnitude planetary nebula *RA 22h 29.6m Dec –20° 48′*
NGC 7009 Saturn Nebula	8th-magnitude planetary nebula *RA 21h 04.2m Dec –11° 22′*
Aries – early winter	
NGC 772	10th-magnitude spiral galaxy *RA 01h 59.3m Dec +19° 01′*
Auriga – winter	
M36 (NGC 1960)	6th-magnitude open cluster *RA 05h 36.1m Dec +34° 08′*
M37 (NGC 2099)	6th-magnitude open cluster *RA 05h 52.4m Dec +32° 33′*
M38 (NGC 1912)	6th-magnitude open cluster *RA 05h 28.7m Dec +35° 50′*
Cancer – late winter to early spring	
M44 (NGC 2632) Praesepe or Beehive	3rd-magnitude open cluster *RA 08h 40.1m Dec +19° 59′*
M67 (NGC 2682)	7th-magnitude open cluster *RA 08h 50.4m Dec +11° 49′*
Canes Venatici – visible all year	
M3 (NGC 5272)	6th-magnitude globular cluster *RA 13h 42.2m Dec +28° 23′*
M51 (NGC 5194/5) Whirlpool Galaxy	8th-magnitude spiral galaxy *RA 13h 29.9m Dec +47° 12′*
M63 (NGC 5055)	9th-magnitude spiral galaxy *RA 13h 15.8m Dec +42° 02′*
M94 (NGC 4736)	8th-magnitude spiral galaxy *RA 12h 50.9m Dec +41° 07′*
M106 (NGC4258)	8th-magnitude spiral galaxy *RA 12h 19.0m Dec +47° 18′*
Canis Major – late winter	
M41 (NGC 2287)	4th-magnitude open cluster *RA 06h 47.0m Dec –20° 44′*
Capricornus – late summer and early autumn	
M30 (NGC 7099)	7th-magnitude globular cluster *RA 21h 40.4m Dec –23° 11′*
Cassiopeia – visible all year	
M52 (NGC 7654)	6th-magnitude open cluster *RA 23h 24.2m Dec +61° 35′*
M103 (NGC 581)	7th-magnitude open cluster *RA 01h 33.2m Dec +60° 42′*
NGC 225	7th-magnitude open cluster *RA 00h 43.4m Dec +61 47′*
NGC 457	6th-magnitude open cluster *RA 01h 19.1m Dec +58° 20′*
NGC 663	Good binocular open cluster *RA 01h 46.0m Dec +61° 15′*
Cepheus – visible all year	
Delta Cephei	Variable star, varying between +3.5 and +4.4 with a period of 5.37 days. It has a magnitude +6.3 companion and they make an attractive pair for small telescopes or binoculars.
Cetus – late autumn	
Mira (omicron Ceti)	Irregular variable star with a period of roughly 330 days and a range between +2.0 and +10.1.
M77 (NGC 1068)	9th-magnitude spiral galaxy *RA 02h 42.7m Dec –00° 01′*

Coma Berenices – spring	
M53 (NGC 5024)	8th-magnitude globular cluster *RA 13h 12.9m Dec +18° 10′*
M64 (NGC 4286) Black Eye Galaxy	8th-magnitude spiral galaxy with a prominent dust lane that is visible in larger telescopes. *RA 12h 56.7m Dec +21° 41′*
M85 (NGC 4382)	9th-magnitude elliptical galaxy *RA 12h 25.4m Dec +18° 11′*
M88 (NGC 4501)	10th-magnitude spiral galaxy *RA 12h 32.0m Dec.+14° 25′*
M91 (NGC 4548)	10th-magnitude spiral galaxy *RA 12h 35.4m Dec +14° 30′*
M98 (NGC 4192)	10th-magnitude spiral galaxy *RA 12h 13.8m Dec +14° 54′*
M99 (NGC 4254)	10th-magnitude spiral galaxy *RA 12h 18.8m Dec +14° 25′*
M100 (NGC 4321)	9th-magnitude spiral galaxy *RA 12h 22.9m Dec +15° 49′*
NGC 4565	10th-magnitude spiral galaxy *RA 12h 36.3m Dec +25° 59′*
Cygnus – late summer and autumn	
Cygnus Rift	Dark cloud just south of Deneb that appears to split the Milky Way in two.
NGC 7000 North America Nebula	A bright nebula against the background of the Milky Way, visible with binoculars under dark skies. *RA 20h 58.8m Dec +44° 20′*
NGC 6992 Veil Nebula (part)	Supernova remnant, visible with binoculars under dark skies. *RA 20h 56.8m Dec +31 28′*
M29 (NGC 6913)	7th-magnitude open cluster *RA 20h 23.9m Dec +36° 32′*
M39 (NGC 7092)	Large 5th-magnitude open cluster *RA 21h 32.2m Dec +48° 26′*
NGC 6826 Blinking Planetary	9th-magnitude planetary nebula *RA 19 44.8m Dec +50° 31′*
Delphinus – late summer	
NGC 6934	9th-magnitude globular cluster *RA 20h 34.2m Dec +07° 24′*
Draco – midsummer	
NGC 6543	9th-magnitude planetary nebula *RA 17h 58.6m Dec +66° 38′*
Gemini – winter	
M35 (NGC 2168)	5th-magnitude open cluster *RA 06h 08.9m Dec +24° 20′*
NGC 2392 Eskimo Nebula	8–10th-magnitude planetary nebula *RA 07h 29.2m Dec +20° 55′*
Hercules – early summer	
M13 (NGC 6205)	6th-magnitude globular cluster *RA 16h 41.7m Dec +36° 28′*
M92 (NGC 6341)	6th-magnitude globular cluster *RA 17h 17.1m Dec +43° 08′*
NGC 6210	9th-magnitude planetary nebula *RA 16h 44.5m Dec +23 49′*
Hydra – early spring	
M48 (NGC 2548)	6th-magnitude open cluster *RA 08h 13.8m Dec –05° 48′*
M68 (NGC 4590)	8th-magnitude globular cluster *RA 12h 39.5m Dec –26° 45′*
M83 (NGC 5236)	8th-magnitude spiral galaxy *RA 13h 37.0m Dec –29° 52′*
NGC 3242 Ghost of Jupiter	9th-magnitude planetary nebula *RA 10h 24.8m Dec –18° 38′*
Leo – spring	
M65 (NGC 3623)	9th-magnitude spiral galaxy *RA 11h 18.9m Dec +13° 05′*
M66 (NGC 3627)	9th-magnitude spiral galaxy *RA 11h 20.2m Dec +12° 59′*
M95 (NGC 3351)	10th-magnitude spiral galaxy *RA 10h 44.0m Dec +11° 42′*
M96 (NGC 3368)	9th-magnitude spiral galaxy *RA 10h 46.8m Dec +11° 49′*
M105 (NGC 3379)	9th-magnitude elliptical galaxy *RA 10h 47.8m Dec +12° 35′*
Lepus – winter	
M79 (NGC 1904)	8th-magnitude globular cluster *RA 05h 24.5m Dec –24° 33′*
Lyra – spring	
M56 (NGC 6779)	8th-magnitude globular cluster *RA 19h 16.6m Dec +30° 11′*
M57 (NGC 6720) Ring Nebula	9th-magnitude planetary nebula *RA 18h 53.6m Dec +33° 02′*
Monoceros – winter	
M50 (NGC 2323)	6th-magnitude open cluster *RA 07h 03.2m Dec –08° 20′*
NGC 2244	Open cluster surrounded by the faint Rosette Nebula, NGC 2237. Visible in binoculars. *RA 06h 32.4m Dec +04° 52′*
Ophiuchus – summer	
M9 (NGC 6333)	8th-magnitude globular cluster *RA 17h 19.2m Dec –18° 31′*
M10 (NGC 6254)	7th-magnitude globular cluster *RA 16h 57.1m Dec –04° 06′*
M12 (NCG 6218)	7th-magnitude globular cluster *RA 16h 47.2m Dec –01° 57′*
M14 (NGC 6402)	8th-magnitude globular cluster *RA 17h 37.6m Dec –03° 15′*
M19 (NGC 6273)	7th-magnitude globular cluster *RA 17h 02.6m Dec –26° 16′*
M62 (NGC 6266)	7th-magnitude globular cluster *RA 17h 01.2m Dec –30° 07′*
M107 (NGC 6171)	8th-magnitude globular cluster *RA 16h 32.5m Dec –13° 03′*
Orion – winter	
M42 (NGC 1976) Orion Nebula	4th-magnitude nebula *RA 05h 35.4m Dec –05° 27′*
M43 (NGC 1982)	5th-magnitude nebula *RA 05h 35.6m Dec –05° 16′*
M78 (NGC 2068)	8th-magnitude nebula *RA 05h 46.7m Dec +00° 03′*
Pegasus – autumn	
M15 (NGC 7078)	6th-magnitude globular cluster *RA 21h 30.0m Dec +12° 10′*
Perseus – autumn to winter	
M34 (NGC 1039)	5th-magnitude open cluster *RA 02h 42.0m Dec +42° 47′*
M76 (NGC 650/1) Little Dumbbell	11th-magnitude planetary nebula *RA 01h 42.4m Dec +51° 34′*

NGC 869/884 Double Cluster	Pair of open star clusters *RA 02h 19.0m Dec +57° 09′* *RA 02h 22.4m Dec +57° 07′*
Pisces – autumn	
M74 (NGC 628)	9th-magnitude spiral galaxy *RA 01h 36.7m Dec +15° 47′*
Puppis – late winter	
M46 (NGC 2437)	6th-magnitude open cluster *RA 07h 41.8m Dec –14° 49′*
M47 (NGC 2422)	4th-magnitude open cluster *RA 07h 36.6m Dec –14° 30′*
M93 (NGC 2447)	6th-magnitude open cluster *RA 07h 44.6m Dec –23° 52′*
Sagitta – late summer	
M71 (NGC 6838)	8th-magnitude globular cluster *RA 19h 53.8m Dec +18° 47′*
Sagittarius – summer	
M8 (NGC 6523) Lagoon Nebula	6th-magnitude nebula *RA 18h 03.8m Dec –24° 23′*
M17 (NGC 6618) Omega Nebula	6th-magnitude nebula *RA 18h 20.8m Dec –16° 11′*
M18 (NGC 6613)	7th-magnitude open cluster *RA 18h 19.9m Dec –17 08′*
M20 (NGC 6514) Trifid Nebula	9th-magnitude nebula *RA 18h 02.3m Dec –23° 02′*
M21 (NGC 6531)	6th-magnitude open cluster *RA 18h 04.6m Dec –22° 30′*
M22 (NGC 6656)	5th-magnitude globular cluster *RA 18h 36.4m Dec –23° 54′*
M23 (NGC 6494)	5th-magnitude open cluster *RA 17h 56.8m Dec –19° 01′*
M24 (NGC 6603)	5th-magnitude open cluster *RA 18h 16.9m Dec –18° 29′*
M25 (IC 4725)	5th-magnitude open cluster *RA 18h 31.6m Dec –19° 15′*
M28 (NGC 6626)	7th-magnitude globular cluster *RA 18h 24.5m Dec –24° 52′*
M54 (NGC 6715)	8th-magnitude globular cluster *RA 18h 55.1m Dec –30° 29′*
M55 (NGC 6809)	7th-magnitude globular cluster *RA 19h 40.0m Dec –30° 58′*
M69 (NGC 6637)	8th-magnitude globular cluster *RA 18h 31.4m Dec –32° 21′*
M70 (NGC 6681)	8th-magnitude globular cluster *RA 18h 43.2m Dec –32° 18′*
M75 (NGC 6864)	9th-magnitude globular cluster *RA 20h 06.1m Dec –21° 55′*
Scorpius (northern part) – midsummer	
M4 (NGC 6121)	6th-magnitude globular cluster *RA 16h 23.6m Dec –26° 32′*
M7 (NGC 6475)	3rd-magnitude open cluster *RA 17h 53.9m Dec –34° 49′*
M80 (NGC 6093)	7th-magnitude globular cluster *RA 16h 17.0m Dec –22° 59′*
Scutum – mid to late summer	
M11 (NGC 6705) Wild Duck Cluster	6th-magnitude open cluster *RA 18h 51.1m Dec –06° 16′*

M26 (NGC 6694)	8th-magnitude open cluster *RA 18h 45.2m Dec –09° 24′*
Serpens – summer	
M5 (NGC 5904)	6th-magnitude globular cluster *RA 15h 18.6m Dec +02° 05′*
M16 (NGC 6611)	6th-magnitude open cluster, surrounded by the Eagle Nebula. *RA 18h 18.8m Dec –13° 47′*
Taurus – winter	
M1 (NGC 1952) Crab Nebula	8th-magnitude supernova remnant *RA 05h 34.5m Dec +22° 00′*
M45 Pleiades	1st-magnitude open cluster, an excellent binocular object. *RA 03h 47.0m Dec +24° 07′*
Triangulum – autumn	
M33 (NGC 598)	6th-magnitude spiral galaxy *RA 01h 33.9m Dec +30° 39′*
Ursa Major – all year	
M81 (NGC 3031)	7th-magnitude spiral galaxy *RA 09h 55.6m Dec +69° 04′*
M82 (NGC 3034)	8th-magnitude starburst galaxy *RA 09h 55.8m Dec +69° 41′*
M97 (NGC 3587) Owl Nebula	12th-magnitude planetary nebula *RA 11h 14.8m Dec +55° 01′*
M101 (NGC 5457)	8th-magnitude spiral galaxy *RA 14h 03.2m Dec +54° 21′*
M108 (NGC 3556)	10th-magnitude spiral galaxy *RA 11h 11.5m Dec +55° 40′*
M109 (NGC 3992)	10th-magnitude spiral galaxy *RA 11h 57.6m Dec +53° 23′*
Virgo – spring	
M49 (NGC 4472)	8th-magnitude elliptical galaxy *RA 12h 29.8m Dec +08° 00′*
M58 (NGC 4579)	10th-magnitude spiral galaxy *RA 12h 37.7m Dec +11° 49′*
M59 (NGC 4621)	10th-magnitude elliptical galaxy *RA 12h 42.0m Dec +11° 39′*
M60 (NGC 4649)	9th-magnitude elliptical galaxy *RA 12h 43.7m Dec +11° 33′*
M61 (NGC 4303)	10th-magnitude spiral galaxy *RA 12h 21.9m Dec +04° 28′*
M84 (NGC 4374)	9th-magnitude elliptical galaxy *RA 12h 25.1m Dec +12° 53′*
M86 (NGC 4406)	9th-magnitude elliptical galaxy *RA 12h 26.2m Dec +12° 57′*
M87 (NGC 4486)	9th-magnitude elliptical galaxy *RA 12h 30.8m Dec +12° 24′*
M89 (NGC 4552)	10th-magnitude elliptical galaxy *RA 12h 35.7m Dec +12° 33′*
M90 (NGC 4569)	9th-magnitude spiral galaxy *RA 12h 36.8m Dec +13° 10′*
M104 (NGC 4594) Sombrero Galaxy	Almost edge-on 8th-magnitude spiral galaxy. *RA 12h 40.0m Dec –11° 37′*
Vulpecula – late summer and autumn	
M27 (NGC 6853) Dumbbell Nebula	8th-magnitude planetary nebula *RA 19h 59.6m Dec +22° 43′*

OBSERVING AT A DISTANCE

Light pollution is a curse to all observers, and many people find that their telescope just doesn't show anything worthwhile from their own gardens because their skies are just too bright. But imaging has given their interest a new lease of life. Deep-sky objects that ceased to become visible years ago now leap into view in seconds on the computer screen. The fun has returned to observing. But the subject has become much more technical as a result.

These days, many telescope mounts and cameras can be controlled directly from a computer, usually a laptop, but increasingly also from smart phones or similar devices. Once you are set up, everything can be done from the keyboard, which is particularly useful if you are into imaging. You choose your target from an on-screen sky map, tell the telescope to slew to it, take an exposure at the click of the mouse and then view the results without having to touch the telescope at all. All this is possible using off-the-shelf components, whether you have a digital SLR (DSLR) camera or a specialist CCD camera.

If you want to carry out a long exposure, requiring careful guiding on a reference star, again you can do this from your keyboard. Modern autoguiders will usually show dozens of stars in any field of view, then it's a simple matter of clicking on one, and the system will lock on to it, allowing for any errors in the telescope drive.

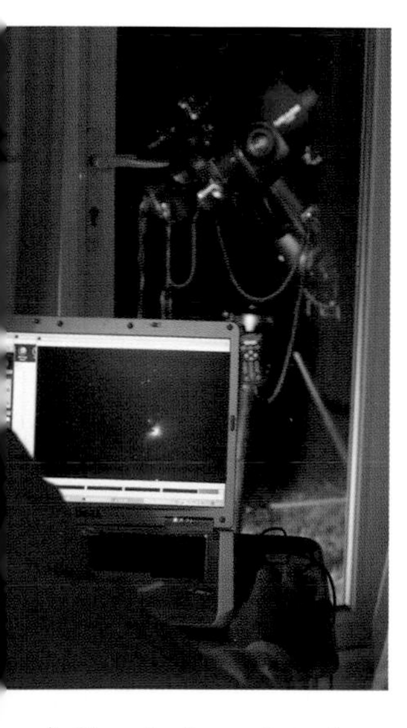

▲ *Observing in comfort using a remotely controlled telescope just requires a laptop and cables long enough to reach the camera and telescope outside.*

The more advanced Go To telescope mounts have a port which will link to your computer, as well as autoguider ports, and are supplied with a suitable connector lead. But be warned – there's more to it than just plug and play. The technology was devised a few years ago, and generally relies on what are called RS-232 or serial connectors to your computer. Now computers have moved on, and have only USB ports these days. So you need to buy a USB-to-serial adapter, which has a particular COM port number depending on which USB socket you plug it into, and you may need to download a driver for your particular telescope from ascom-standards.org. Then you'll need to configure your software for your scope and the COM port for your adapter, making sure you always plug the adapter into the same USB socket. This all sounds a bit of a minefield, but unfortunately most telescope manufacturers leave it up to the customer to find their way through it all!

There are many advantages to controlling your telescope in this way, not least that if your leads are long enough you can now retreat indoors with your computer while leaving your telescope out in the cold where it works best. Air turbulence caused by temperature differences around or inside the telescope spoils images, so observing from indoors is not usually feasible. By linking your telescope to a camera and computer, you can now enjoy your virtual observing in comfort, even from suburban areas.

A standard DSLR camera will take reasonable images of the brighter deep-sky objects within a few seconds, though for the best results you need to take several, indeed many, subexposures and combine them in editing software such as Photoshop or Paint Shop Pro. You can even overcome the worst effects of light pollution, which otherwise turns the sky a very unexciting muddy brown. Using a CCD camera and autoguider – which each may cost about the same or more than a DSLR – you can image surprisingly faint objects even with small telescopes. But that's another story.

▲ *The iTelescope.net observatory at Siding Spring Mountain, New South Wales, Australia, lies alongside large professional telescopes. Shown here are 90 mm and 106 mm refractors, plus 318 mm, 431 mm and 510 mm reflectors.*

REMOTE OBSERVING

If you can observe with your telescope some distance away from you, the next step is to automate the whole thing so that you don't have to go outside at all. Put the scope in a dome with motorized roof, make sure you can control the scope, camera, focus and all the rest of it from your keyboard – oh, and while you're at it, put the telescope much farther away in a dark-sky site! Some enterprising amateurs have done this, but for most of us the effort is too great. However, in the past few years remote observing facilities have been set up which allow other users to control the telescope and camera via the Internet.

Obviously such remote observing systems aren't generally free to use, though they are set up by keen amateurs so there isn't a great profit motive in there. The telescopes are set up in dark-sky sites and are generally at the upper end of quality or sizes used by amateurs, in some cases overlapping with instruments that professionals might use. They are equipped with cameras and filters that most amateurs would love to own, and are maintained in full working order.

A leading provider, iTelescope.net, has over a dozen telescopes, ranging in aperture from 90 mm up to half a metre, spread across the world at Siding Spring Observatory in Australia, Nerpio in Spain and

▼ *A 4-minute exposure through a hydrogen-alpha filter taken using the 510 mm reflector at Siding Spring initially shows only a pale image of the object.*

▲ *The final result, using a number of separate images taken through different filters, shows the Eagle Nebula and the 'Pillars of Creation' made famous by the Hubble Space Telescope image of the same area.*

Mayhill in New Mexico, USA. This network means that at least one site is in darkness at any time, and with sites in both hemispheres it's possible to image objects that are invisible from your own location. Imagine taking photos with a half-metre telescope in real time in your lunch break at your desk, or even in a coffee bar! Other providers include Sierra Stars Observatory Network, with telescopes up to 0.81-metre aperture in the western USA; Myastropic, which currently has telescopes in Spain; and Lightbuckets, with telescopes in France.

You don't need to wait around for a telescope to be free, however. You can reserve observing time in advance, write an exposure plan, and leave the telescope to take the shots automatically. If it's cloudy or the dome has to be closed for some other reason, you will receive a credit and can schedule the shots for some other time.

Some of the smaller telescopes have basic single-shot colour cameras, which will give you a good result straight away. The basic exposure time suggested is 5 minutes, which is fine for the brighter objects, but fainter ones will need considerably longer. The night's main deep-sky objects are listed for you, so a single click is all you need. You can view the telescope's progress of focusing and star-finding by means of a continually updated script, and a preview image is emailed to you immediately afterwards.

The full-size image, corrected for the known characteristics of the instrumentation, is delivered to a folder which you can access over the web or using FTP software. File sizes can be large (typically 11 Mb per shot, even when zip compressed), so you need a good broadband connection. The primary image format of the images is FITS, which is not handled by most image viewers, though you can also request a TIFF file, which is more common.

Most of the telescopes have mono cameras, which are more sensitive than colour cameras, and allow greater flexibility through the use of various colour filters. To get the greatest sensitivity, you choose the Luminosity option, which is basically unfiltered. For colour results, you need to take separate exposures through colour filters – either red, green and blue for a basic colour image, or hydrogen alpha, oxygen or sulphur to bring out features in particular objects.

Though the iTelescope.net website offers useful videos which will help you to use the instruments, the choice of filters and exposure times is left up to you, which can be daunting for beginners. You may find that you need to buy image-processing software to get the best out of the images.

WEBSITES

Remote observing facilities mentioned in this article:

www.iTelescope.net

www.sierrastars.com

myastropic.com

lightbuckets.com

THE BOTTOM LINE

So what's the cost? Typically you buy points or credits in advance to use on any of the telescopes. This covers the actual exposure time only – all the focusing and slewing to the object is not charged for. Each telescope has its own rate per imaging hour, which may be lower when there is moonlight. A 30-minute exposure, yielding a colour image of a bright object, could cost you around £15 on a small telescope to £60 or more on a large one (though the fainter objects may require exposures of many hours). Rates vary depending on the service provider, but they are all broadly similar. iTelescope.net offer a free trial run on their 150 mm refractor in New Mexico, to give you the feel of the experience.

These prices might seem steep for a single picture, but consider that buying your own equipment to do the same thing would cost more than £2000 for even the most basic equipment, and you'll still have to battle with the local climate and maybe light pollution. For that sum you could take over 100 pictures with much better equipment under ideal dark-sky conditions. And also consider what other activities or hobbies might cost: a dinner out, or a trip to see your favourite musical performer or sports event, for example, could set you back much more.

Of course, acquiring good images is not everything in amateur astronomy. The photo you take from your back garden using what equipment you have can be much more satisfying than any taken remotely. There are many parallels – would you prefer to photograph your local beauty spot in person or view the Grand Canyon on a remote-controlled webcam? Or catch a glimpse of an unusual bird in your back garden rather than see one on TV?

Nevertheless, remote observing is becoming popular even among amateur astronomers with their own instruments. There are limitations – currently, most systems are designed for long-exposure photography through telescopes, and are not capable of imaging the Moon or planets, or provide wide-field views of constellations, for example. But there's a lot to be said for being able to take a great astrophoto from your desk on a rainy day in the middle of winter!

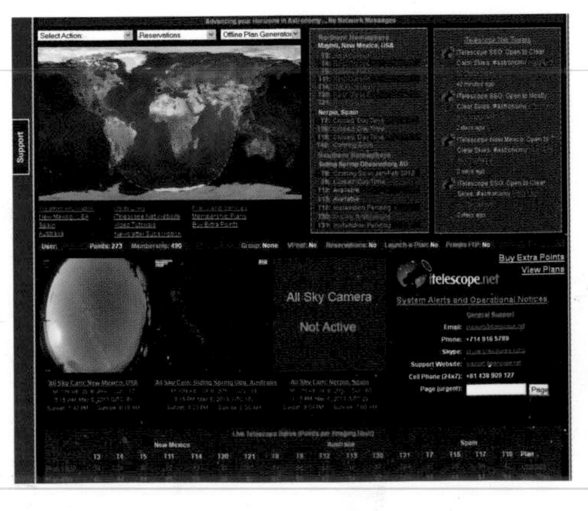

▼ *From this web page you can see the observing conditions at three sites around the world and the availability of the instruments. Clicking on a free telescope puts you in control of it.*